Innovative Materials in Sustainable Civil Infrastructure

Innovative Materials in Sustainable Civil Infrastructure

Edited by

Dr. Richa Palod

ARCLER PRESS

www.arclerpress.com

Innovative Materials in Sustainable Civil Infrastructure

Dr. Richa Palod

Arcler Press
4164 Lakeshore Road
Burlington, ON L7L 1A4
Canada
www.arclerpress.com
Email: orders@arclereducation.com

ISBN: 978-1-77956-433-7 (Hardcover)

ABOUT THE EDITOR

Dr. Richa Palod, has completed her Doctoral research in Structural Engineering, a Master's in Structural Engineering, and a Bachelor's in Civil Engineering. She has more than 8 years of teaching experience working with learners of different age groups. After completing her Ph.D. in Structural Engineering, she taught students spanning across continents, ranging from Asia and Australia to the United Kingdom. She has published articles related to innovative construction materials in SCI/Scopus indexed journals. She is also a reviewer and editor for Q1 journals indexed in SCI/Scopus. Her commitment to staying informed about innovations in the construction and civil engineering sectors is ongoing, and you can find further on www.civiljournals.com.

TABLE OF CONTENTS

LIST OF FIGURES

LIST OF TABLES

PREFACE

Globally, environmentally sustainable construction materials with high performance have been in great demand by the construction industry. So far now concrete and steel are the most popular man-made materials for construction in the world. Despite numerous great infrastructure projects around the world, many of them are gradually deteriorating. It is expected that infrastructure repair costs in the world will approach trillions of dollars over the next twenty years. One of the key causes of decay in our infrastructures is the deterioration of the materials used in their development and repair.

Concrete is brittle and consequently at risk of cracking, whilst structural steel is highly susceptible to corrosion. These serious infrastructural issues highlight the urgent need for more durable and innovative construction materials that will eventually replace traditional concrete and steel and will most likely replace these popular materials soon.

The discussion of the books starts with the introduction to sustainable civil infrastructure and materials, offering a comprehensive exploration of the evolution, current trends, and prospects in the field. From ancient construction methods to cutting-edge, fiber reinforced polymers, advanced concretes, recycled aggregates and fiber reinforced composite materials.

This book comprises of eight chapters; the first chapter discusses the fundamental principles of sustainable development and tracing the evolution of materials in civil engineering. The second chapter discusses a diverse array of innovative materials, from advanced concretes to smart materials.

Chapter 3 focuses on the rehabilitation of infrastructure using advanced polymer composites, showcasing their applications, characteristics, and manufacturing processes. While the fourth chapter explores the utilization of recycled aggregates in geotechnical applications, shedding light on their properties, performance, and environmental impact.

Chapter 5 describes the advanced concrete from its high strength to eco-friendly variants, and their numerous applications in civil infrastructure. Chapter 6 explores deeper into advanced cementitious composites, exploring their mechanical properties, manufacturing processes, and applications in structural engineering. Chapter 7 shifts the focus to fiber-reinforced composites, examining their advantages, challenges, and applications in civil infrastructure projects. From enhancing structural integrity to minimizing life-cycle costs, and the eighth chapter explores advanced steel applications in civil engineering, addressing issues such as fire protection, fatigue, and corrosion resistance. This chapter demonstrates the ongoing evolution of steel as a vital component of sustainable infrastructure.

Through the collective expertise of contributors from academia, industry, and research institutions, this book endeavors to provide a comprehensive overview of innovative materials in sustainable civil infrastructure. It is our hope that this volume will serve as a valuable resource for engineers, researchers, and policymaker's alike, inspiring continued innovation and collaboration in the pursuit of a more sustainable built environment.

CHAPTER 1

INTRODUCTION TO SUSTAINABLE CIVIL INFRASTRUCTURE AND MATERIALS

LEARNING OBJECTIVES

At the end of this chapter, readers will be able to:

1. Understand the principles of sustainable development in civil infrastructure.
2. Learn about the historical evolution of civil engineering materials.
3. Understand the significance of innovative materials in sustainable construction practices.
4. Comprehend the impact of ancient construction materials on modern engineering.
5. Understand the role of modern advancements in civil engineering materials.
6. Understand future trends shaping the field of civil engineering.
7. Gain knowledge of materials science to optimize infrastructure sustainability.

INTRODUCTORY EXAMPLE

Turner Construction Company, a prominent leader in sustainable construction practices, undertook a significant renovation project converting a historic building into a modern office space in downtown Chicago. This project exemplifies their commitment to integrating innovative materials and sustainable principles into civil infrastructure. Turner Construction opted to utilize cutting-edge sustainable materials such as recycled steel and concrete with high fly ash content sourced locally. These materials not only reduced the project's carbon footprint but also promoted circular economy principles by minimizing waste and supporting local suppliers.

In addition to material choices, Turner implemented state-of-the-art energy-efficient systems. They installed advanced composite insulating panels made from recycled materials for superior thermal performance. Furthermore, the building features smart sensors and controls that optimize energy consumption by adjusting lighting and HVAC systems based on occupancy and environmental conditions. Through these innovations, Turner Construction achieved significant reductions in energy usage and operational costs while enhancing indoor environmental quality. This project underscores their role in advancing sustainable civil infrastructure practices, demonstrating how innovative materials and technologies can be seamlessly integrated to achieve both environmental stewardship and operational excellence. This example showcases Turner Construction Company's leadership in adopting sustainable materials and practices, highlighting their proactive approach in transforming traditional construction methods towards more sustainable and resilient infrastructure solutions.

UNIT INTRODUCTION

Within this chapter, the key concepts of environmental sustainability and the progressive path towards materials in civil engineering are explained (Mirza, 2006). Modern engineering initiatives are guided by the principles of sustainable development, which consider economic, environmental, and social factors harmoniously. Delving into its principles, a deep understanding of how sustainability goes beyond simple conservation efforts, influencing every aspect of infrastructure development, design, and implementation (Thorpe, 2015). Moreover, developments of materials in civil engineering, a journey that stretches from ancient times to the latest advancements of the twenty-first century are described in this chapter. With a rich history of progress and ingenuity, the narrative of civil engineering materials showcases a continuous drive for innovation and flexibility in response to the ever-changing demands of society (Shu et al., 2013).

Within this chapter, the importance of materials within civil engineering is explained, examining their profound influence on the constructed surroundings. Exploring the evolution of construction materials, from the resources of the past to the impressive structures of today, reveals the profound impact these materials have had on shaping our architectural world (Zhou et al., 2017).

1.1. UNDERSTANDING SUSTAINABLE DEVELOPMENT

Learning Objectives

- Understand the role of innovative materials in advancing sustainable civil infrastructure.
- Comprehend the common characteristics and challenges of infrastructure assets.

Sustainable development is an urgent requirement in today's world, encompassing the integration of economic progress, social equality, and environmental conservation. This signifies a fundamental change in the understanding of progress, highlighting the interconnectedness of human welfare, environmental sustainability, and fairness across generations. The key to sustainable development is the idea of combining different elements, acknowledging their interconnectedness, and planning to create a more stable and adaptable future (Boeren, 2019) (Figure 1.1).

Figure 1.1. *Illustration of sustainable development innovation*

Source: Naida Ademovic, Creative Commons License.

1.1.1. Definition and Principles

Sustainable development embraces a comprehensive approach to development, acknowledging the intricate interconnections between economic prosperity, social equity, and environmental well-being. Emphasizing the importance of integration, it promotes solutions that effectively meet the needs of current and future generations. The foundation of its viewpoint revolves around the idea of interdependence, highlighting the interconnections between human societies and the natural world, and the importance of considering the long-term consequences of our impact on the planet along its inhabitants (Mensah, 2019).

1.1.2. Importance in Civil Infrastructure

In the field of civil infrastructure, the concept of sustainable development acts as a guiding light, showing the way to resilient, resource-efficient, adaptable systems of infrastructure. Here, the incorporation of innovative materials becomes a crucial approach to promoting sustainable practices. Exploring new materials and technologies can greatly improve the performance and lifespan of civil infrastructure, all while reducing its impact on the environment (Hollaway, 2010). From sustainable concrete mixes to energy from renewable sources systems, these innovations hold great potential. In addition, engineers are encouraged to take a comprehensive approach to infrastructure projects, considering not only their functional needs but also their wider effects on ecosystems, communities, and economies, in line with sustainable development principles (Bolar et al., 2014). Through the adoption of sustainable practices and the utilization of innovative materials, civil infrastructure has the potential to go beyond its conventional function and play a pivotal role in driving positive transformations, ultimately contributing to the achievement of a more sustainable and fair society (Blagojević et al., 2022).

Infrastructure assets, including energy utilities, roads, railways, bridges, pipelines, and residential and commercial buildings share several similarities. They play a crucial role in supporting the essential services that enable contemporary societies and worldwide society to operate smoothly. These structures are intended to respond to the unique needs of their respective regions and are considered durable assets that can endure for many generations (Fischer & Amekudzi, 2011). Certain types of infrastructure, like transportation, power, and water systems, depend on intricate networks to fulfill their functions and can gain advantages via economies of scope and scale. Despite these challenges, infrastructure services continue to play a crucial role in fulfilling the needs of society. The infrastructure currently developing has the potential to address the evolving needs and challenges of society. By considering the increasing constraints and higher expectations, ensuring that this infrastructure provides safety, security, and necessary adaptations to tackle climate change, population growth, and the protection of the planet's support systems (Frangopol & Liu, 2019).

1.2. EVOLUTION OF MATERIALS IN CIVIL ENGINEERING

Learning Objectives

- Understand the significance and characteristics of civil engineering materials.
- Understand the historical evolution and sustainability of ancient construction materials.
- Know the role of ancient materials like mud and lime in early civilizations.

Gaining insight into the historical development of materials used in civil engineering is crucial in the modern era, as there is an increasing focus on creating infrastructure that is both sustainable and resilient. Having a deep understanding is crucial for engineers to make informed decisions that combine traditional knowledge with modern technologies, ensuring the durability of structures in the long run.

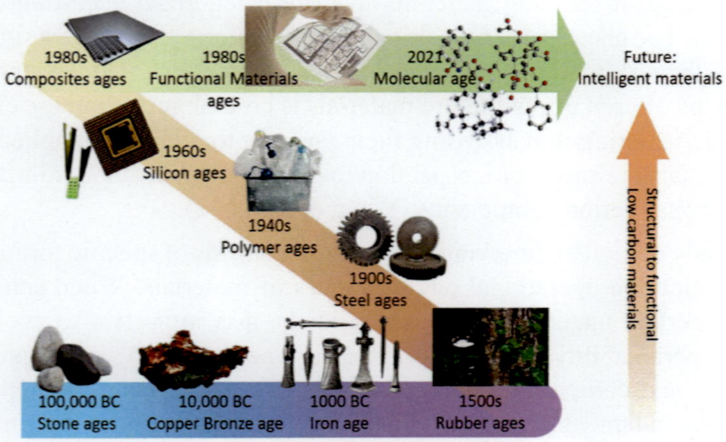

Figure 1.2. *Illustration of historical and emerging tendencies of engineering materials.*

Source: Li Jingcheng, Creative Commons License.

Appreciating the importance of cultural heritage, this understanding encourages the maintenance of historic structures as well as methods in construction procedures (Poullain et al., 2006) (Figure 1.2). Firstly, it provides a context for the current state of construction materials, aiding engineers and researchers in comprehending the advancements and developments.

In addition, it shares valuable insights gained from past successes and failures of materials, contributing to the progress of contemporary materials and construction methods. Gaining a deeper understanding of the durability and sustainability of materials as time passes involves examining the choices made by ancient civilizations and the consequences that followed (Bundur et al., 2017).

1.2.1. Significance of Civil Engineering Materials

Engineering places great emphasis on selecting the appropriate materials and equipment to effectively apply its fundamental principles. Civil engineering encompasses more than just technical analysis and complex structural understanding. It also emphasizes the importance of a civil engineer's mindset in determining the appropriate tools and strategies for successful planning and implementation. Understanding the properties and advancements of specific materials is essential when incorporating them into research, as it plays a vital role in project planning (Rao et al., 2015). Every building material has its distinct composition, durability, classifications, and properties. There is a diverse range of building materials that are categorized according to their components. The chemical composition of these materials can have a significant impact on their suitability for use in various applications. Understanding the compositions of civil infrastructure materials is crucial, including the existence of inorganic materials and assessing their capacity to withstand applied forces. These materials are subject to inspection to determine their standard when used in various construction components (Wang et al., 2022).

The study of civil engineering involves the analysis of specific formulas that help determine the operational circumstances of materials. When considering the use of certain materials under heavy loads, it is important to avoid those with high permeability properties. Here is where a deep understanding of calculations and comprehension becomes crucial to choosing the appropriate materials. A fundamental principle often observed in civil engineering materials is their ability to withstand external forces, requiring careful evaluation to accurately determine their characteristics. Various states of stress that can be applied to a material include its ability to withstand tension, compression, bending, and shear forces. Considering the susceptibility to deformation is an important factor in understanding the importance of building materials (Schabowicz, 2021). Under significant loads, a material has the potential to undergo deformation based on its inherent strength. The material's brittleness is defined by its tendency to fracture under stress, and it is also associated with deformation. External forces or vibrations can have a significant impact on the toughness of civil infrastructure, potentially leading to deformation. Understanding the characteristics of building materials is crucial, as a faulty load on one component can render a structure ineffective. Establishing a solid foundation is crucial for every project plan, which also encompasses the considerations of material selection.

Over time, extensive research and study have been dedicated to understanding and refining the fundamental principles underlying the construction of various building materials. This ongoing pursuit has shed light on their essential functions and significance (Afzal et al., 2020).

1.2.2. Ancient Construction Materials

Even after thousands of years of weathering and use, certain old building materials are still in use today in constructions. For instance, the Roman remains cover a wide range of climates and seismic zones, extending from England to the Middle East. The Romans employed locally obtained materials as the raw components for concrete, which they used as their main building material. The mortar that has endured for thousands of years was produced at temperatures much lower than those used to make Portland cement today, and it has a demonstrated lifespan longer than the infrastructure's planned life. Archaeologists specializing in the study of past cultures have found that ancient civilizations often built their homes using materials readily available in their surroundings.

Tree branches were commonly utilized as makeshift structures against natural formations like cliffs, rocks, or other supports to construct lean-tos. In addition, huts were built by skillfully intertwining tree branches to form sturdy frames and ingeniously utilizing animal skins to create portable structures that closely resembled tents (Yang et al., 2017). Throughout history, mud has been widely utilized as a fundamental building material in ancient construction. Mud was utilized in different ways, either in its natural state for wall building or in the form of blocks of mud or bricks. Additionally, it played a crucial role in connecting the building blocks and providing a smooth surface for wall plastering. It was also commonly used as a durable flooring material. The essential components include spirit, meticulously trimmed straw, and water. The components are carefully combined through the process of shoveling and compacting, resulting in a mixture that is then shaped into bricks of a standardized size using an open mold (Oleson et al., 2014). (Figures 1.3 and 1.4).

Figure 1.3. *Illustration of Original raw materials found in the privernum archaeological site and used in buildings and decorations. Wall paintings with original pigment fragments (a), various types of milled ceramic (b) and sand (c), fragment of lime (d) and bricks and stones for floor decorations.*

Source: Alfred Swenson, Creative Commons License.

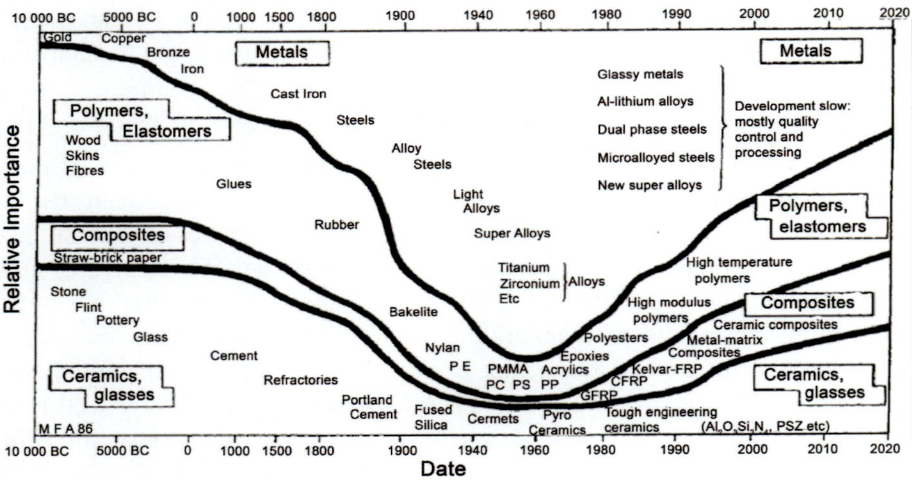

Figure 1.4. *Illustration of the evolution of the engineering materials.*

Source: André Canal Marques, Creative Commons License.

Mud buildings have proven to be more than just historically significant. They offer numerous advantages, such as their ability to effectively regulate temperature and their sustainable nature. The flexibility of this product enables it to be used in various ways, while its ability to be recycled addresses environmental considerations. Mud constructions provide a practical solution to address the challenges of global warming, offering natural cooling properties and improved air circulation. Nevertheless, it is crucial to recognize that dealing with mud can require a significant amount of physical effort (Khitab & Anwar, 2016).

Mortar, a time-honored building material, has played a vital role in the connection of masonry elements like stones and bricks. In ancient times, lime or lime-based mortar was the preferred choice for mortar. This type of mortar was made from quicklime, which was produced by heating limestone at high temperatures. Throughout history, this material has been a crucial component in construction, both in ancient times and in modern applications, where it interacts seamlessly with cement. Lime is a readily available construction material that can be handled rapidly into a usable state, resulting in the construction of much more substantial and strong buildings (Flaga, 2000). The longest wall of China, built between 221 BCE as well 1644 CE, was predominantly constructed with stone or clay-fired bricks that were bonded together using lime binders. Typically, it does not include sand-to-gravel-sized aggregate. Instead, it is made up of hydrated quicklime paste. Lime has been widely utilized in strengthening the foundations of civil infrastructure, providing support to pillars and roofs. The durability of ancient Roman structures was enhanced using self-healing concrete that incorporated a reactive chemical form of lime called "quicklime." The exceptional characteristics of the ancient Roman infrastructure played a significant role in its long-lasting and robust nature. This specific type of lime does not exist in contemporary concrete (Pachta et al., 2014).

1.3. INDUSTRIAL REVOLUTION AND NEW MATERIALS

Learning Objectives

- Understand the historical evolution and significance of construction materials.
- Comprehend the impact of industrialization on construction technology and methods.
- Understand the advancements like Reinforced Concrete (RC), steel alloys, and nanotechnology.

The 18th century was the beginning of the industrial era, which led to substantial advancements in industry. Although there were no significant breakthroughs in construction methods during this era, the 19th century saw remarkable progress in developing construction materials. With the advent of new materials such as cast iron, wrought iron, and eventually steel, a whole new world of possibilities opened in the field of construction. This allowed for the creation of innovative structures like building frames, railways, and bridges. The use of glass became essential in constructing steel-framed buildings featuring expansive glazed envelopes. Additionally, the introduction of Portland cement played a significant role in the development of concrete and, subsequently, reinforced concrete structures (Rymarczyk, 2020).

During the later part of the 19th century, a burgeoning industry arose that specialized in manufacturing various building equipment, including elevators, pipes, radiators, boilers, and the Industrial Revolution. Then along with the need for additional housing in Europe after World Wars I and II, particularly the latter, provided the foundation for the advancement of more effective construction methods. This required a transition from conventional labor-intensive techniques to contemporary approaches, commonly known as industrialization of construction (Dobrzański & Dobrzańska-Danikiewicz, 2019).

There have been different proposed definitions of construction industrialization, with some focusing on mechanization and differentiating it based on construction technology rather than the actual product. Sebestyen explains that industrialization in construction entails the implementation of innovative technologies, which include in situ procedures like slip-forms over chimneys, and silos. This industrialization is characterized by modern design methods incorporating scientific information about structures, building physics, fire, and the use of computers (Bradu et al., 2023).

Prefabrication, as a method of industrialization, entails the manufacturing of building components in a factory setting, away from the construction site. This method can be used for a wide range of structural materials, such as steel, polymers, timber, concrete, and aluminum. Prefabrication has been widely utilized in the construction of various types of buildings, including houses and multi-story industrial structures. These include portal houses, mobile homes, manufactured housing, and system buildings. Although prefabrication has achieved notable accomplishments, it has also encountered criticism from certain experts who have expressed concerns about its effectiveness (Kurtuluş et al., 2020).

Mechanization plays a vital role in the progress of construction technology, contributing to the overall development of industrialization. The introduction of the engine with internal combustion (gasoline or petroleum) and the electrical motor revolutionized the construction industry, expanding its possibilities beyond what the steam engine could offer. In the past, the functioning components of construction machines relied on ropes for operation. However, modern advancements have led to the adoption of hydraulic, mechanical, pneumatic, and electronic systems, or a combination of these, to drive these parts (Wilkinson, 1988). Various construction machinery with varying sizes and capacities have been developed to cater to different job requirements. In recent years, the construction industry has also welcomed the integration of robots into various processes, including excavation, tunneling, moving, slip-form procedures, and underwater work. Manufacturing construction machinery has given significant industrialized countries an important advantage in the global market. This sector, along with contracting, consulting, and materials for construction production, is crucial to the construction industry (Jemghili et al., 2020).

1.3.1. Concrete and Modern Building Materials

Over the years, concrete has undergone significant transformations, evolving from its early forms to the sophisticated cement-based material used in contemporary construction. The history of concrete can be traced back to its origins, with significant contributions from ancient civilizations such as the Egyptians, Romans, and the Greeks. Vitruvius, the renowned Roman architect, extensively documented the utilization of pozzolanic components within concrete. His invaluable contributions shed light on the early compositions of concrete, offering profound insights into this crucial aspect of civil infrastructure. The Industrial Revolution that took place in the 19th century marked a pivotal turning point in the development of concrete. In the year 1824, Joseph Aspdin revolutionized the construction industry with his invention of Portland cement. This remarkable innovation paved the way for the modern manufacturing of concrete, rivaling the high standard of building stone found in Portland, England (Kumar, 2012).

Reinforced concrete, a groundbreaking combination of steel and concrete, came into prominence during the latter part of the 19th and beginning of the 20th century. The innovative application of reinforced concrete by Francois Hennebique revolutionized construction practices, enabling the creation of taller and exceptionally resilient buildings. Advancements in concrete technology have continued to evolve in the twenty-first century, with a strong emphasis on sustainability and improving performance. Emerging technologies in the field of concrete structures are revolutionizing their strength, durability, and environmental sustainability (Bautista-Gutierrez et al., 2019).

The importance of glass in construction has evolved from being a functional material to becoming a crucial element in contemporary architectural design. The introduction of float glass in the mid-20th century revolutionized the field, providing architects and engineers with transparent and undistorted material for windows and facades. Smart glass and energy-efficient glazing systems, for instance, play a significant role in promoting sustainability and enhancing the comfort of occupants in contemporary buildings.

Aluminum's lightweight and corrosion-resistant properties have solidified its position as a prominent material in contemporary construction. Aluminum alloys gained significant popularity in the mid-20th century, leading to their increased usage in structural components, roofing, and facades. The malleability and recyclability of aluminum contribute to the promotion of sustainable construction methods (Meyer, 2005) (Figure 1.5).

Figure 1.5. *Illustration of the modern building materials.*

Source: Eva Nie, Creative Commons License.

The historical evolution of concrete, coming from ancient compounds to modern cement-based combinations, showcases its remarkable versatility and enduring significance in the field of construction. In addition, the incorporation of reinforced concrete, glass, and aluminum has revolutionized the construction industry, offering architects and engineers a plethora of innovative and environmentally friendly design possibilities.

Thorough ongoing materials science research and development, significant advancements are being made to revolutionize the construction industry. These innovations are leading to the creation of more durable, efficient, and environmentally conscious building solutions, shaping the future of civil infrastructure (Dmitrieva et al., 2018).

Practice Problem

"GRT Construction Limited specializes in delivering high-quality, innovative building solutions", has been assigned the responsibility of building a sustainable housing complex, with a strong focus on using environmentally friendly materials and implementing sustainable practices. Designing a concrete mix that meets a 40 MPa compressive strength demand at 28 days poses a challenge. The mix incorporates 30% recycled aggregate, utilizes Portland cement Type I, maintains a water-cement ratio of 0.45, adds 20% fly ash as a partial substitute for cement, and reduces the amount of water by 15% using a high-range water-reducing admixture.

Solutions to Practice Problem

To tackle this issue, GRT Construction Limited takes great care in maintaining a delicate balance of material proportions. The amount of recycled aggregate needed is calculated and the amount of cementitious materials needed per volume of concrete is determined. After adjusting the water content for the admixture, the weights of the fine and coarse aggregates are then calculated. This guarantees a durable concrete mix that fulfills strength criteria, incorporates recycled materials and supplementary cementitious materials, and minimizes environmental impact, in line with the environmentally conscious objectives of the project.

1.3.1. Innovation and Advancement in the 20th Century

In the 20th century, several technological advancements greatly transformed the materials used in civil engineering, leading to significant changes in construction practices. Reinforced concrete (RC) is a composite material that is widely utilized in the construction of buildings and civil infrastructures. It stands as a prime example of an innovative solution in the field. Reinforcing steel is incorporated into concrete to withstand various forces. In 2012, worldwide manufacturing of reinforced concrete reached a staggering annual volume of over 10 billion cubic meters. This massive production not only consumed a great deal of natural resources but also had significant environmental consequences. Steel and concrete reinforcement are the primary components that contribute to the overall weight of RC structures. In recent decades, there have been significant efforts to optimize different types of RC structures and components, aiming to minimize the need for steel and concrete reinforcement (Best, 2019).

In addition, advancements in metallurgy have resulted in the development of high-strength metal alloys, including steel alloys and high-strength steel. Due to their enhanced tensile strength and flexibility, such alloys played a crucial role in the construction of high-rise buildings, bridges, and various other large-scale structures. In addition, the field of materials science has seen the integration of nanotechnology, allowing for precise manipulation of materials at the molecular and atomic scales. This breakthrough has resulted in the creation of construction materials that are not only stronger and lighter but also more durable. Examples of such materials include nanomaterial-enhanced concrete and coatings. Scientists have explored different approaches to incorporate nanomaterials into composite materials to enhance their properties and structural performance (Harman, 1980). Furthermore, Fiber-reinforced Polymers (FRP) find extensive use in new construction as well as repaired applications, effectively addressing numerous limitations associated with conventional construction materials. FRP materials offer excellent resistance to electro-chemical corrosion and possess impressive strength-to-weight and stiffness-to-weight ratios. Moreover, their installation flexibility allows for relatively easy installation in repair applications using limited access. Consequently, considered a reliable and durable option for the restoration of worn-out concrete structures. This includes reinforcing and fixing flexural issues, enhancing shear strength, and providing column support (Oki, 2019) (Figure 1.6).

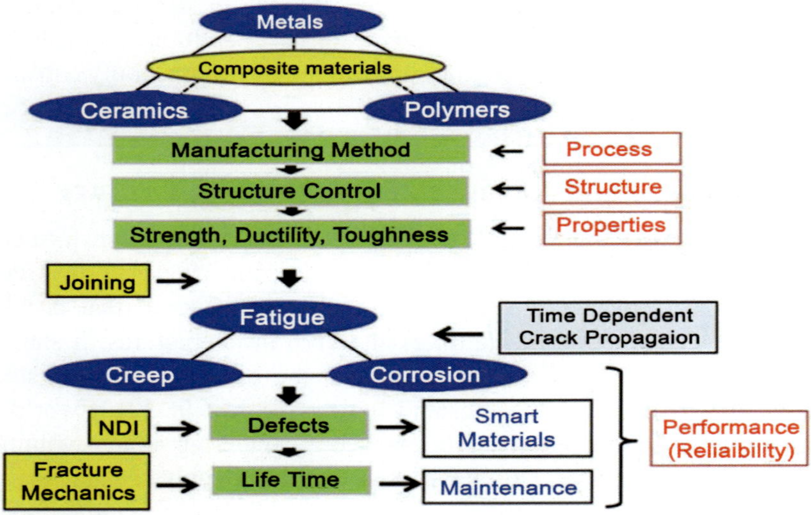

Figure 1.6. *Illustration of the development of innovative structural materials.*

Source: Robert Hull, Creative Commons License.

1.4. FUTURE TRENDS

Learning Objectives

- Understand the impact of nanotechnology on enhancing concrete properties.
- Understand the integration of innovative technologies in civil engineering trends.

In a world characterized by constant change, civil engineering stands at the forefront of adapting and innovating, addressing the dynamic needs of our continuously evolving built environment. This explores the emerging trends that are transforming the field of civil engineering, ushering in a new era of infrastructure that prioritizes not only safety but also enhanced efficiency, sustainability, and resilience. These advancements strive to fulfill the needs of present and future generations, demonstrating a dedication to long-lasting and forward-thinking infrastructure (Salehi et al., 2021).

Nano-chemistry presents groundbreaking products that can be applied in the field of concrete technology. As an example, the development of innovative concrete admixtures, such as polycarboxylic ether superplasticizers, focuses on extending the time that concrete mixtures maintain their desired consistency. Integrating nanoparticles into conventional building materials can significantly improve the characteristics, rendering them ideal for the construction of advanced civil infrastructure such as high-rise buildings, long-span structures, and smart infrastructure. As an illustration, nanoparticles of are utilized as additives within high-performance and self-compacting concrete, resulting in improved workability and strength. The application of nanotechnology in concrete presents a promising opportunity for the construction industry. It provides the potential to develop building materials that are not only more durable and sustainable but also capable of meeting the evolving demands of modern infrastructure (Balaji et al., 2022).

The rise of biodegradable building materials has the potential to greatly impact sustainability initiatives by providing a practical solution to the environmental issues associated with conventional construction materials such as cement. Through the utilization of agricultural waste, this emerging practice not only helps reduce pollution caused by traditional materials but also addresses the issue of waste disposal within landfills. Exploring the utilization of agricultural by-products such as peanut shells, rice straw, rice husks, and coconut shells. The incorporation of agricultural residues as a partial replacement for sand in the manufacturing of cement blocks led to the development of blocks that successfully met the rigorous ASTM standards for strength and durability. This groundbreaking approach not only showcases the practicality of utilizing agricultural waste but also emphasizes its potential in producing construction materials that adhere to industry standards, promoting a more sustainable and ecologically conscious construction environment (Sandanayake et al., 2020).

The current trends in infrastructure development encompass the integration of advanced technologies, creative construction methods, environmentally friendly design approaches, and a holistic view of building civil infrastructure. An in-depth analysis of these trends offers helpful insights into the continuous development of civil engineering and highlights its crucial role in driving transformative changes. Multifaceted development strives to create a global environment that is not just sustainable and resilient, but also responsive to the needs of the future. Exploring the intricate details of these trends reveals the complex web of advancements that are shaping the field of civil engineering. These advancements are making a significant contribution to the development of an improved, sustainable, efficient, and built environment for the future (Hossain, 2015).

SUMMARY

- The narrative underscores the evolution of civil engineering, from ancient construction materials to modern innovations. Sustainable development principles guide infrastructure advancements, integrating economic, social, and environmental considerations.
- Notably, the Industrial Revolution spurred the introduction of materials like steel and concrete, while contemporary trends emphasize sustainability and resilience. Nanochemistry enhances concrete properties, while biodegradable materials offer eco-friendly alternatives.
- These trends signify a holistic approach to infrastructure development, blending cutting-edge technologies with sustainable design principles. Understanding the historical evolution of materials informs modern engineering practices, emphasizing durability and cultural preservation.
- Ultimately, civil engineering adapts to dynamic demands, shaping a future of resilient, efficient, and sustainable infrastructure.

MULTIPLE CHOICE QUESTIONS

1. **What does sustainable development emphasize in civil engineering?**
   a. Economic prosperity only
   b. Environmental preservation only
   c. Social equity only
   d. Environmental preservation, social equity, and economic prosperity
2. **Which subsection of Understanding Sustainable Development discusses the core concepts that drive sustainable practices in infrastructure development?**
   a. Definition and Principles
   b. Importance in Civil Infrastructure
   c. Significance of Civil Engineering Materials
   d. Innovation and Advancement in the 20th Century
3. **Why is understanding sustainable development crucial for civil engineers?**
   a. It's not crucial
   b. It helps in maximizing profits

c. It ensures infrastructure meets the needs of the present without compromising future generations

d. It speeds up construction projects

3. Which era witnessed significant advancements in material science and engineering, leading to the emergence of new materials in civil engineering?

a. Medieval times

b. Renaissance period

c. Industrial Revolution

d. Stone Age

4. What is considered a quintessential emblem of modern civil engineering in the 20th century?

a. Mud bricks

b. Wooden beams

c. Concrete

d. Straw bales

5. Which subsection of the chapter discusses the transformative impact of the Industrial Revolution on civil engineering materials?

a. Definition and Principles

b. Importance in Civil Infrastructure

c. Concrete and Modern Building Materials

d. Innovation and Advancement in the 20th Century

REVIEW QUESTIONS

1. What are the core principles of sustainable development, and how do they apply to civil infrastructure?
2. Why is understanding sustainable development crucial for civil engineers? Provide examples to support your answer.
3. Discuss the significance of civil engineering materials in infrastructure projects. How does material selection affect the overall sustainability and functionality of built structures?
4. Briefly describe the evolution of civil engineering materials from ancient times to the modern era. Highlight key advancements and their impact on construction practices.

5. How did the Industrial Revolution contribute to the advancement of civil engineering materials? Provide examples of new materials and technologies introduced during this period.
6. Explain the importance of concrete as a modern building material. What are its properties and how does it influence contemporary construction practices?

Answers to Multiple Questions

1. (d)   2. (a)   3. (c)   4. (c)   5. (c)   6. (d)

REFERENCES

1. Afzal, A., Kausar, A., & Siddiq, M. (2020). Role of polymeric composite in civil engineering applications: A review. *Polymer-plastics Technology and Materials, 59*(10), 1023–1040. https://doi.org/10.1080/03602559.2020.1812745.
2. Balaji, C. R., de Azevedo, A. R., & Madurwar, M. (2022). Sustainable perspective of ancillary construction materials in infrastructure industry: An overview. *Journal of Cleaner Production, 365*, 132864. https://doi.org/10.1016/j.jclepro.2022.132864.
3. Best, M. H. (2019). Industrial innovation and productive structures: The creation of America's 'Arsenal of democracy.' *Structural Change and Economic Dynamics, 48*, 32–41. https://doi.org/10.1016/j.strueco.2018.12.005.
4. Bonen, D., Tasdemir, M. A., & Sarkar, S. (1994). The evolution of cementitious materials through history. *MRS Online Proceedings Library (OPL), 370*, 159–161. https://doi.org/10.1557/PROC-370-159.
5. Bradu, P., Biswas, A., Nair, C., Sreevalsakumar, S., Patil, M., Kannampuzha, S., & Gopalakrishnan, A. V. (2023). Recent advances in green technology and Industrial Revolution 4.0 for a sustainable future. *Environmental Science and Pollution Research, 30*(60), 124488–124519. https://doi.org/10.1007/s11356-023-25961-7.
6. Dmitrieva, M., Leitsin, V., Sharanova, A., & Tovpinets, A. (2018). Complex approach to the study of modern concrete characteristics. *MATEC Web of Conferences, 251*, 1024–1028. https://doi.org/10.1051/matecconf/201825110024.
7. Edgerton, D. (2004). 'The linear model' did not exist: Reflections on the history and historiography of science and research in industry in the twentieth century. In *The Science-Industry Nexus: History, Policy,*

Implications, 7(3), 31–57. https://doi.org/10.1007/s12053-004-0028-7.

8. Fischer, J. M., & Amekudzi, A. (2011). Quality of life, sustainable civil infrastructure, and sustainable development: Strategically expanding choice. *Journal of Urban Planning and Development, 137*(1), 39–48. https://doi.org/10.1061/(ASCE)UP.1943-5444.0000046.
9. Flaga, K. (2000). Advances in materials applied in civil engineering. *Journal of Materials Processing Technology, 106*(1–3), 173–183. https://doi.org/10.1016/S0924-0136(00)00623-1.
10. Frangopol, D. M., & Liu, M. (2019). Maintenance and management of civil infrastructure based on condition, safety, optimization, and life-cycle cost. *Structures and Infrastructure Systems, 7*(3), 96–108. https://doi.org/10.1080/15732479.2010.531246.
11. Hall, M. R., Dehdezi, P. K., Dawson, A. R., Grenfell, J., & Isola, R. (2012). Influence of the thermophysical properties of pavement materials on the evolution of temperature depth profiles in different climatic regions. *Journal of Materials in Civil Engineering, 24*(1), 32–47. https://doi.org/10.1061/(ASCE)MT.1943-5533.0000386.
12. Harman, A. J. (1980). Industrial innovation and governmental policy: A review and proposal based on observation of the US electronic sector. *Technological Forecasting and Social Change, 18*(1), 15–37. https://doi.org/10.1016/0040-1625(80)90027-0.
13. Hossain, K. M. (2015). Sustainable infrastructure development strategies through technological innovation of smart materials. *Journal of Emerging Trends in Engineering and Applied Sciences, 6*(7), 180–187. https://doi.org/10.1109/ICECOCS.2020.9295860.
14. Jemghili, R., Taleb, A. A., & Mansouri, K. (2020). Additive manufacturing progress as a new industrial revolution. In *2020 IEEE 2nd International Conference on Electronics, Control, Optimization and Computer Science (ICECOCS)* (Vol. 7, pp. 1–8). IEEE. https://doi.org/10.1109/ICECOCS51217.2020.9379496.
15. Khitab, A., & Anwar, W. (2016). Classical building materials. In *Advanced Research on Nanotechnology for Civil Engineering Applications* (Vol. 3, pp. 1–27). IGI Global. https://doi.org/10.4018/978-1-4666-9427-5.ch001.
16. Kumar, R. S. (2012). Experimental study on the properties of concrete made with alternate construction materials. *International Journal of Modern Engineering Research, 2*(5), 3006–3012. https://doi.org/10.1007/s12053-013-9253-4.
17. Kurtuluş, M., Güner, S., & Arpacıoğlu, Ü. T. (2020). Changes in the use of materials in the field of architecture after the industrial revolution. In *XIV*

International Theory and History of Architecture Conference November (Vol. 6, pp. 249–256). https://doi.org/10.1016/j.matpr.2020.04.194.

18. Mensah, J. (2019). Sustainable development: Meaning, history, principles, pillars, and implications for human action: Literature review. *Cogent Social Sciences, 5*(1), 1653531. https://doi.org/10.1080/23311886.2019.1653531.
19. Meyer, C. (2005). Concrete as a green building material. In *Construction Materials Mindess Symposium* (Vol. 10, pp. 94–99). https://doi.org/10.1111/j.1468-0327.2006.00147.x.
20. Mirza, S. (2006). Durability and sustainability of infrastructure—A state-of-the-art report. *Canadian Journal of Civil Engineering, 33*(6), 639–649. https://doi.org/10.1139/l06-005.
21. Oleson, J. P., Jackson, M. D., Hohlfelder, R. L., & Brandon, C. J. (2014). Building for eternity: The history and technology of Roman concrete engineering in the sea. *Building for Eternity, 1*, 350–368. https://doi.org/10.1515/9783110337398.350.
22. Pachta, V., Stefanidou, M., Konopisi, S., & Papayianni, I. (2014). Technological evolution of historic structural mortars. *Journal of Civil Engineering and Architecture, 8*(7), 987–990. https://doi.org/10.17265/1934-7359/2014.07.001.
23. Poullain, P., Mounanga, P., Bastian, G., & Coue, R. (2006). Determination of the thermophysical properties of evolutive porous media: Application to civil engineering materials. *The European Physical Journal-Applied Physics, 33*(1), 35–49. https://doi.org/10.1051/epjap:2006060.
24. Rao, N. V., Rajasekhar, M., Vijayalakshmi, K., & Vamshykrishna, M. (2015). The future of civil engineering with the influence and impact of nanotechnology on properties of materials. *Procedia Materials Science, 10*, 111–115. https://doi.org/10.1016/j.mspro.2015.06.022.
25. Rymarczyk, J. (2020). Technologies, opportunities and challenges of the industrial revolution 4.0: Theoretical considerations. *Entrepreneurial Business and Economics Review, 8*(1), 185–198. https://doi.org/10.15678/EBER.2020.080111.
26. Sandanayake, M., Bouras, Y., Haigh, R., & Vrcelj, Z. (2020). Current sustainable trends of using waste materials in concrete—A decade review. *Sustainability, 12*(22), 9622. https://doi.org/10.3390/su12229622.
27. Schabowicz, K. (2021). Testing of materials and elements in civil engineering. *Materials, 14*(12), 3412. https://doi.org/10.3390/ma14123412.

28. Shu, X., Huang, B., & Liu, J. (2013). Special issue on materials innovations for sustainable infrastructure. *Journal of Materials in Civil Engineering, 25*(7), 825–828. https://doi.org/10.1061/(ASCE)MT.1943-5533.0000732.
29. Wilkinson, R. G. (1988). The English industrial revolution. In *The Ends of the Earth: Perspectives on Modern Environmental History* (pp. 741–745). https://doi.org/10.1017/CBO9780511815838.034.

CHAPTER 2

TYPES OF INNOVATIVE MATERIALS

LEARNING OBJECTIVES

At the end of this chapter, readers will be able to:

1. Understand the significance of sustainable materials in civil infrastructure.
2. Understand the key trends driving material innovation for infrastructure.
3. Comprehend the characteristics and applications of advanced concretes.
4. Know the benefits and uses of recycled aggregates in construction.
5. Understand bio-based aggregates and their potential in sustainable construction.
6. Know the properties and applications of Fiber-Reinforced Polymers (FRP).
7. Gain insights into the role of smart materials in enhancing infrastructure performance.

INTRODUCTORY EXAMPLE

Acme Engineering, a company renowned for its commitment to pushing the boundaries of construction technology. In a recent project, Acme was tasked with designing a bridge to connect two lively urban centers. Instead of resorting to conventional materials, they opted for a blend of advanced concretes and sustainable aggregates. This decision not only ensured the bridge's structural integrity but also significantly reduced its carbon footprint, aligning with the city's sustainability goals.

Through strategic partnerships with material innovators, such as EcoTech Solutions and SmartBuild Technologies, Acme infused the bridge with smart materials capable of monitoring structural health in real-time. This proactive approach to maintenance not only enhances safety but also minimizes downtime and repair costs over the bridge's lifespan.

Acme's dedication to practical applications of innovative materials isn't limited to bridges. Their portfolio boasts a diverse range of projects, from eco-friendly skyscrapers to resilient coastal defenses, all united by a common thread: the relentless pursuit of sustainable, high-performance construction solutions.

UNIT INTRODUCTION

Novel materials are transforming the civil infrastructure sector, providing answers to enduring problems while driving the construction sector towards resilient and sustainable development. This module investigates the wide range of materials that are transforming the way infrastructure is developed and maintained. Advanced concretes featuring self-healing qualities, sustainable aggregates, and new composites each offer distinct advantages and uses in the construction industry. In addition, the use of intelligent materials allows for continuous monitoring and proactive maintenance, hence improving the efficiency and durability of infrastructure systems. This course explores the potential of modern building materials to revolutionize civil infrastructure and shape the future. It focuses on cutting-edge substances and their applications (Abyzov et al., 2020).

2.1. INNOVATIVE MATERIALS IN CIVIL INFRASTRUCTURE

Learning Objectives

- Understand the importance of sustainable materials in addressing environmental challenges.
- Explore key trends in material innovation for sustainable infrastructure solutions.

2.1.1. Importance of Sustainable Materials

Building materials prioritizing sustainability are crucial in tackling environmental issues and advancing the construction industry towards a more sustainable future. With the growth of the global population and urbanization, there is an increasing need for buildings and infrastructure. Nevertheless, the rapid increase in construction has had a detrimental impact on the environment, resulting in the depletion of resources, pollution, and the release of greenhouse gases. Within the current setting, sustainable building materials present an intriguing solution that not only lessens the environmental consequences of construction but also enhances the effectiveness, longevity, and well-being of buildings (Akadiri, 2015).

The significance of sustainable building materials lies in their ability to contribute to environmental preservation. Conventional construction practices have long been dependent on materials that require significant resources, including concrete, steel, and timber. When it comes to sustainable building materials, the focus is on utilizing low-carbon or carbon-neutral options. These include industrial hemp, straw bales, reclaimed wood, and bamboo. Utilizing these cutting-edge materials, building initiatives can greatly diminish their environmental impact, as these materials demand less energy and generate less greenhouse gas emissions through the manufacturing process. Building materials derived from industrial hemp promote the effective utilization of resources. Furthermore, it has the potential to be quickly replenished, offering viable alternatives to resources that have slow growth rates or are non-renewable (Gaustad et al., 2011).

2.1.2. Trends in Material Innovation for Infrastructure

Current developments in the materials industry involve a growing emphasis on sustainability, as materials are being designed to be easily recycled and have a minimal impact on the environment. There is a growing trend in various industries to seek out materials that are both lightweight and durable, to enhance efficiency and performance. This trend is especially evident in the aerospace and automotive sectors (Hossain, 2015). There is a growing movement towards the use of biodegradable materials and the adoption of bio-based alternatives. Companies are embracing environmentally friendly practices and meeting consumer expectations by utilizing bio-based materials. These materials offer a practical solution to the widespread problem of waste, particularly in the packaging industry (Kurpayanidi & Makhmudova, 2016).

Innovative materials, like self-healing polymers, recycled aggregates, fiber reinforced polymers, and smart materials, are offering solutions to some of the most urgent problems of our time. Consider the aerospace industry as an example, where the incorporation of cutting-edge composites is revolutionizing the field of aircraft design. It enhances the lightweight nature of aircraft, resulting in enhanced performance and fuel efficiency (Kurtis, 2015).

2.2. TYPES OF INNOVATIVE MATERIALS IN INFRASTRUCTURE

Learning Objectives

- Understand the development and applications of advanced concretes in infrastructure.
- Understand the characteristics and applications of sustainable aggregates in construction.
- Know about novel composites, including fiber-reinforced polymers and bamboo-reinforced concrete.
- Gain insights into smart materials, such as shape memory alloys and piezoelectric materials, in civil engineering.

2.2.1. Advanced Concretes

Concrete is an ever-evolving material, with the definitions of long-lasting and sustainable concrete evolving. The long-term viability of cementitious building materials has emerged as a crucial subject in the field of concrete science in recent times. Concrete structures necessitate advanced materials that possess desirable characteristics. There have been recent advancements in the development of new types of concrete that focus on improving the microstructure. These advancements involve using alternative binders instead of typical Portland cement and incorporating modern ad-hoc materials and activators. The main goal of these innovations is to enhance the durability, safety, and general ecological impact of concrete (Makul, 2020).

2.2.1.1. Self-healing Concrete

Revolutionary advancements have been made in the realm of self-healing materials. Concrete is a unique material that possesses the remarkable ability to repair itself. The self-healing or self-repairing ability of concrete is a remarkable feature that allows it to mend itself as cracks appear. The strength of concrete represents the key characteristic that sets it apart in its ability to endure physical and chemical harm. There is a clear relationship between a reduction in durability along an increased likelihood of cracking. Various types of cracks can occur in concrete, including settlement fractures, shear, flexural cracks, plastic shrinkage cracks, expansion and cracks, and a combination of these. Various factors contribute to the issue, including splits due to reinforcing corrosion, poor construction, heat strains, and cracks related to weathering (Gardner et al., 2018) (Figure 2.1).

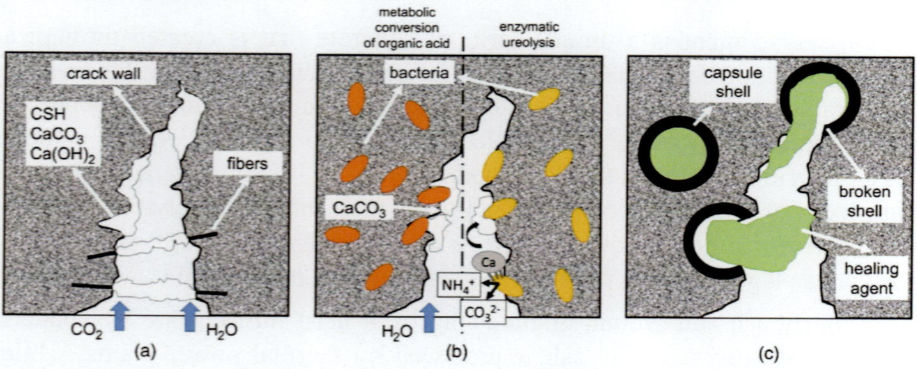

Figure 2.1. *Illustration of self-healing mechanism in self-healing concrete.*

Source: Yohgesh Morthy, Creative Commons License.

2.2.1.2. Ultra-High-Performance Concrete (UHPC)

UHPC is a cementitious material known for its exceptional strength and durability. This can become a practical solution for enhancing the long-term viability of construction sites and other components of infrastructure. UHPC has gained significant traction in numerous countries in recent years. Its applications span a wide range, including repair, rehabilitation, architectural features, building components, and bridges, along with vertical structures like windmill towers and utility towers. Additionally, it finds use in the oil and gas industry, offshore structures, hydraulic structures, and overlay materials. Road and bridge building are widely recognized as the primary applications for UHPC. There have been numerous types of UHPC that have been developed by different manufacturers and in different countries. These include CERACEM, BSI, compact reinforced composites (CRC), multi-scale cement composite (MSCC), and reactive powder concrete (RPC). Across the globe, there are numerous examples of remarkable achievements in the utilization of UHPC. Nevertheless, there are still challenges that restrict its implementation. Continued research and exploration are addressing areas of uncertainty to pave the way for the future adoption of groundbreaking, cost-effective, environmentally friendly, practical, and financially viable Ultra-High-Performance Concrete (UHPC), which will greatly enhance its widespread acceptance. This paper offers a comprehensive overview of UHPC, including its definitions, applications, development, and challenges. It presents the latest information on this innovative material (Nodehi & Nodehi, 2022).

2.1.2.3. Geopolymer Concrete

Geopolymer cement is a unique form of concrete that is created through a chemical reaction between specific materials and a caustic activator. This results in a durable and environmentally friendly alternative to traditional concrete. It has the potential to serve as a viable alternative to regular Portland cement (OPC). Davidovits coined the term "geopolymer" in 1978 to describe materials that consist of chains or networks of inorganic molecules. Geopolymer concrete with cement is produced by utilizing waste materials like fly ash and ground-granulated slag from blast furnaces (GGBS) (Wasim et al., 2021).

Both fly ash and ground granule slag from blast furnaces are byproducts of industrial processes. Fly ash is produced by thermal power plants, while ground aggregate blast furnace ash is a waste material from steel plants. These include fly ash and GGBS undergo advanced processing techniques and are utilized in the construction industry as geopolymer concrete. This concrete solution offers a sustainable approach by effectively minimizing waste stock and carbon emissions through a reduction in the demand for Portland cement. Geopolymers are primarily composed of silicon and aluminum, which are derived from thermally activated organic substances like kaolinite or industrial byproducts such as fly ash or slag. These components are combined with an alkaline activating solution, which causes them to polymerize into molecular chains or networks, resulting in a durable binder. It is sometimes referred to as alkali-activated cement and inorganic polymer cement (Bellum et al., 2020).

2.2.2. Sustainable Aggregates

Aggregates are unprocessed materials obtained from natural sources and collected from pits and quarries. They include crushed stone, gravel, and sand. When combined with a binding agent, such as water, cement, and asphalt, they are utilized to create composite materials, like asphalt concrete and Portland cement concrete. Aggregates are the primary volume of concrete and possess a naturally low carbon footprint. The majority of these materials are naturally found and require no processing. They are typically obtained from nearby sources, which helps to reduce carbon emissions related to transportation (Hu et al., 2019).

2.2.2.1. Recycled Aggregates

Aggregates are particulate substances utilized in construction activities. The materials encompass recycled asphalt, recycled concrete, recycled brick, and slags. Recycled aggregate includes materials such as recycled concrete made from construction and demolition waste, as well as railway ballast. Asphalt pavement is obtained from roadways, parking lots, and runways, while cement concrete might be obtained from bridges, walkways, and roads. After being gathered and treated, these reclaimed aggregates can be utilized in the building of bike lanes, road shoulders, and several other construction applications (Han et al., 2023) (Figure 2.2).

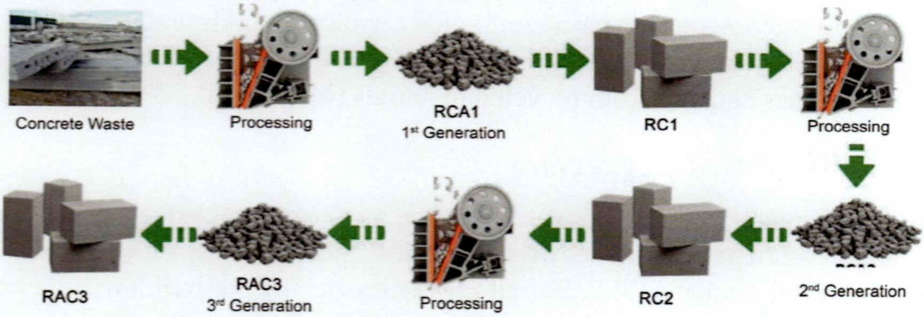

Figure 2.2. *Illustration of multi recycled aggregates.*

Source: Angel Salesa, Creative Commons License.

> **Practice Problem**
>
> SANYO Constructions is a prominent construction company known for its expertise in delivering high-quality infrastructure and building projects, recently the SANYO was assigned a task to increase the sustainability of its concrete production by incorporating recycled aggregates into its mix. They need to determine the percentage of recycled aggregates required to replace 30% of the traditional aggregates in a 50 cubic meter concrete batch. Given the varying densities of traditional and recycled aggregates, SANYO Constructions seeks a solution to optimize the use of recycled materials while maintaining the quality and performance of their concrete products.

Solutions to Practice Problem

SANYO Constructions decides to calculate the percentage of recycled aggregates needed based on volume rather than mass, considering the different densities of the aggregates. After careful analysis, they find that approximately 37.5% of the total volume of the concrete batch needs to be composed of recycled aggregates to replace 30% of the traditional aggregates. This solution ensures a sustainable approach to concrete production while meeting quality standards and reducing environmental impact, aligning with SANYO Constructions commitment to sustainability and innovation in material sourcing.

Using recycled materials during construction is a forward-thinking approach within the construction industry. There are several advantages to using this approach, such as minimizing the reliance on new materials, decreasing energy usage, minimizing waste sent to landfills, and reducing emissions. Using aggregate materials not only benefits the environment but also offers cost savings and the potential to earn credits for certification. Around two hundred million tons for aggregates are utilized annually in the UK, with approximately 57 million tones sourced from recycled materials (Mistri et al., 2020).

2.2.2.2. Bio-Based Aggregates

Plant-based materials, known as bio-aggregates, are utilized as substitutes for mineral aggregates obtained from quarries. Incorporating a Bio-aggregate can significantly enhance the desired characteristics of the building material. Neighborhood Construction is examining the feasibility, effectiveness, and agricultural aspects of several plant-based materials. These options consist of Oil Seed Rape (OSR) straw, which is a byproduct of food production, or Miscanthus Giganteus, a high-yield crop that may be cultivated on marginal terrain that is not appropriate for food production. Cannabis Sativa, sometimes known as industrial Hemp, is widely recognized and frequently utilized as a Bio-aggregate (Kazemi et al., 2022).

2.2.3. Novel Composites

Composites can be created using a wide variety of materials, such as wood and other plant-based substances, which can be transformed into strands, veneer, lumber, fibers, and particles (Hollaway, 2010).

2.2.3.1. Fiber-Reinforced Polymers (FRP)

Fiber-Reinforced Polymer (FRP), also known as Fiber-reinforced plastics, is a form of composite that combines a polymer matrix with fibers to create a strong and durable material. Typically, the fibers used are glass, aramid, and carbon, but occasionally alternative fibers like paper, wood, or asbestos have been utilized. Typically, a thermosetting plastic such as epoxy, vinyl ester, or polyester is used as the phenol-formaldehyde resins and polymer, remaining to be utilized. FRPs are widely utilized in various industries such as marine, aerospace, automotive, and construction. Composite materials are created by combining two or more materials that have distinct properties, resulting in a structure that maintains the individual characteristics of each component. Many composites consist of robust and rigid fibers embedded in a matrix that is comparatively weaker and less rigid. The goal is typically to create a component that possesses strength and rigidity, often while maintaining a low density. FRP composites are utilized in a wide range of advanced engineering structures. These structures include aircraft, helicopters, spacecraft, boats, ships, automobiles, chemical processing equipment, offshore drilling rigs, and civil infrastructure like bridges and buildings (Casadei et al., 2005).

FRPs are becoming more and more popular to improve or replace infrastructure components made of traditional materials like concrete and steel. FRP composites possess numerous advantages, such as their lightweight nature, resistance to high specific strength, corrosion, and stiffness, ease of construction, and ability to be customized to meet specific performance needs. FRP composites have proven to be highly beneficial in various construction applications. And widely adopted in new construction and rehabilitation projects, where used as formwork, modular structures, reinforcement in concrete, and bridge decks, for external reinforcement to enhance strength and seismic resistance (Fang et al., 2019) (Figure 2.3).

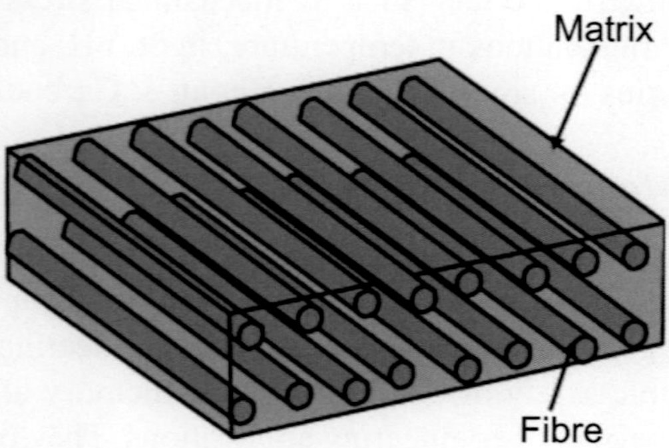

Figure 2.3. *A schematic diagram showing a typical unidirectional FRP plate.*

Source: Abu-Farsakh, Creative Commons License.

2.2.3.2. Bamboo-Reinforced Concrete

Bamboo is a highly versatile plant that finds extensive use in various industries, agriculture, and construction. Over the years, bamboo has garnered significant attention in the construction industry because of its eco-friendly nature, long-lasting durability, and impressive strength. The incorporation of bamboo as a reinforcement in concrete structures represents a cutting-edge advancement in the utilization of bamboo within the construction industry. Using bamboo as reinforcement in concrete structures provides a range of benefits. Bamboo reinforced concrete offers a sustainable solution to traditional steel supported concrete by significantly reducing energy requirements throughout its lifecycle, including manufacturing, transportation, raw material acquisition, usage, and recycling/disposal (Amin et al., 2021). Bamboo has been utilized as a reinforcement within concrete to enhance its strength and endurance. BRC has completed numerous building projects on a global scale, encompassing various types of structures such as residential buildings, bridges, and towering skyscrapers. Using bamboo for structural reinforcement over concrete offers numerous advantages over traditional steel reinforcement. Furthermore, bamboo is a readily accessible and sustainable resource that can be found in various regions across the globe (Karthik et al., 2018).

2.2.4. Smart Materials

Smart construction materials, also referred to as intelligent materials or adaptive materials, possess the ability to effectively and deliberately react to alterations in their surroundings or conditions. Various factors can trigger alterations in the properties of smart materials, such as mechanical stress/strain, electrical/magnetic fields, or fluctuations in temperature, light, pH, and moisture. There are different categories for smart materials (Chong & Garboczi, 2002).

2.2.4.1. Shape Memory Alloys (SMAs)

These materials have the remarkable ability to return to their original size and shape when exposed to specific temperature changes. Shape memory alloys are being increasingly used in the field of civil engineering, particularly for enhancing the seismic safety of buildings. Shape memory alloys have proven to be highly effective in civil engineering applications. They possess the unique ability to absorb strain energy repeatedly without undergoing permanent deformation. Hence, becomes ideal for withstanding a wide range of cyclic behavior and providing excellent fatigue resistance, under large strain cycles. Their exceptional durability and reliability make them a valuable long-term solution (Zareie et al., 2020) (Figures 2.4 and 2.5).

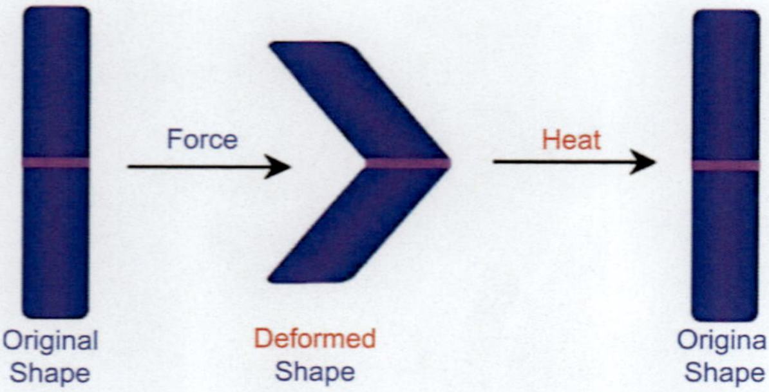

Figure 2.4. *Description of smart alloy materials working principle.*

Source: Milad Bazil Creative Commons License.

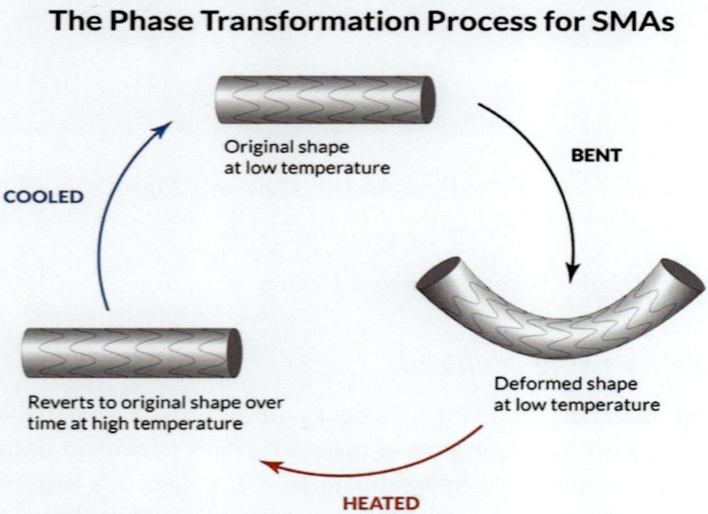

Figure 2.5. *Illustration of the phase transformation of the SMA's.*

Source: Robert Lopez, Creative Commons License.

2.2.4.2. Piezoelectric Materials

These materials can generate voltage as surface strain is applied. On the other hand, the materials experience deformation (stress) whenever an electric field is applied to them. When incorporated into a structural component, a piezoelectric material produces an electric field in reaction to mechanical forces (Song et al., 2017) (Figure 2.6).

Figure 2.6. *Illustration of the Piezoelectric Materials Illustration of the Phase transformation of the SMA's.*

Source: Daniel P Lok, Creative Commons License.

2.2.4.3. Self-sensing Materials

Self-sensing materials, a unique category of electronic materials, can be developed by modifying their nano or microstructures to convert desired states into detectable or noticeable transformations. For instance, a material can be enhanced with a piezoelectric conduct to convert mechanical stress through electrical charges. It can also be modified via mechano-chromatic photon crystals to produce a mechano-chromatic reaction to mechanical strain. Additionally, it can be functionalized using electrical properties to convert mechanical strain into a detectable change at an electrical signal. The article primarily focuses on the most recent family of electrical methods for Structural Health Monitoring (SHM) due to their widespread popularity. These approaches are favored for SHM applications because of their scalability, simplicity of implementation, and direct correlation between signal and damage (Wang et al., 2024).

2.3. APPLICATIONS OF INNOVATIVE MATERIALS IN INFRASTRUCTURE

Learning Objectives

- Explore the applications of innovative materials in bridges and highways.
- Explore the applications of innovative materials in Buildings and Structures

2.3.1. Bridges and Highways

The use of cutting-edge materials has completely transformed the way bridges and highways are built and maintained, providing improved longevity, environmental friendliness, and economic efficiency. Advanced composites, like Fiber-Reinforced Polymers (FRPs), offer exceptional strength-to-weight ratios, corrosion resistance, and decreased maintenance needs. These qualities make them highly suitable for bridge decks and structural elements. Novel materials with self-healing properties have revolutionized the durability of concrete and asphalt, providing exceptional resistance to cracks and potholes. This breakthrough technology significantly extends the lifespan of highways and minimizes the necessity for frequent repairs. Moreover, the integration of cutting-edge materials equipped with sensors allows for the continuous monitoring of structural integrity, resulting in improved safety measures and the implementation of proactive maintenance approaches for transportation infrastructure (Leshko, 2006).

2.3.2. Buildings and Structures

The use of different types of concrete, such as fiber-reinforced concrete and geopolymer concrete, represents a significant breakthrough in the field of construction and structural materials. Utilizing cutting-edge materials in academic writing allows for the exploration of new possibilities and advancements. Fiber-reinforced concrete, when combined with steel, glass, or synthetic fibers, showcases remarkable improvements in tensile strength, crack resistance, and ductility. This breakthrough enables the creation of structures that are not only resilient but also long-lasting. Geopolymer concrete is a highly durable and chemically resistant alternative to traditional Portland cement-based concrete. It is made from chemical byproducts or natural materials, making it an environmentally friendly choice with a reduced carbon footprint. These cutting-edge concrete formulations not only enhance structural performance but also promote sustainable construction practices, creating a more environmentally friendly and durable built environment (Zolotov et al., 2021).

SUMMARY

- The chapter probes into the pivotal role of advanced materials in revolutionizing civil infrastructure. It underscores the importance of sustainable materials in construction, highlighting trends driving material innovation.
- Exploring various types of innovative materials, such as advanced concretes, sustainable aggregates, novel composites, and smart materials, the chapter provides insights into their unique properties and applications.
- From self-healing concrete to shape memory alloys, these materials offer solutions for enhancing the durability, sustainability, and functionality of infrastructure projects.
- The chapter concludes by examining the diverse applications of innovative materials in bridges, highways, buildings, and other structures, displaying their transformative impact on modern construction practices.

MULTIPLE CHOICE QUESTIONS

1. **Which type of innovative material is designed to autonomously repair cracks in its structure?**
    a. Recycled Aggregates
    b. Self-Healing Concrete
    c. Shape Memory Alloys (SMAs)
    d. Bamboo-reinforced Concrete
2. **What type of concrete is known for its exceptional strength and durability, allowing for thinner and more resource-efficient structures?**
    a. Ultra-High-Performance Concrete (UHPC)
    b. Geopolymer Concrete
    c. Fiber-Reinforced Polymers (FRP)
    d. Bio-based Aggregates
3. **Which innovative material utilizes natural fibers to enhance the properties of concrete?**
    a. Piezoelectric Materials
    b. Shape Memory Alloys (SMAs)
    c. Bamboo-reinforced Concrete
    d. Self-Sensing Materials

4. **What type of material is capable of changing its shape in response to external stimuli, such as temperature or stress?**
   a.. Ultra-High Performance Concrete (UHPC)
   b. Recycled Aggregates
   c. Shape Memory Alloys (SMAs)
   d. Fiber-Reinforced Polymers (FRP)
5. **Which innovative material can generate an electrical charge in response to mechanical stress, making it suitable for energy harvesting applications?**
   a. Piezoelectric Materials
   b. Self-Healing Concrete
   c. Recycled Aggregates
   d. Self-Sensing Materials

REVIEW QUESTIONS

1. What is the significance of sustainable materials in civil infrastructure?
2. Describe the key features of Ultra-HighPerformance Concrete (UHPC) and its potential applications.
3. How do recycled aggregates contribute to sustainability in construction projects?
4. Compare Fiber-Reinforced Polymers (FRP) and Bamboo-reinforced Concrete in terms of their properties and applications.
5. Explain the function of Shape Memory Alloys (SMAs) in infrastructure and provide a real-world example of their use.

Answers to Multiple Questions

1. (b) 2. (a) 3. (c) 4. (c) 5. (a)

REFERENCES

1. Abyzov, V. A., Pushkarova, K. K., Kochevykh, M. O., Honchar, O. A., & Bazeliuk, N. L. (2020). Innovative building materials in creation an architectural environment. In *IOP Conference Series: Materials Science and Engineering* (Vol. 907, pp. 12035–12040). IOP Publishing. https://doi.org/10.1088/1757-899X/907/12/12035.

2. Akadiri, P. O. (2015). Understanding barriers affecting the selection of sustainable materials in building projects. *Journal of Building Engineering, 4*, 86–93. https://doi.org/10.1016/j.jobe.2015.08.001.
3. Amin, K. F., Asrafuzzaman, Sharif, A., & Hoque, M. E. (2021). Bamboo/bamboo fiber reinforced concrete composites and their applications in modern infrastructure. *Bamboo Fiber Composites: Processing, Properties and Applications, 4*(3), 271–297. https://doi.org/10.1016/j.jobe.2021.102383.
4. Bellum, R. R., Muniraj, K., & Madduru, S. R. C. (2020). Characteristic evaluation of geopolymer concrete for the development of road network: Sustainable infrastructure. *Innovative Infrastructure Solutions, 5*(3), 91. https://doi.org/10.1007/s40940-020-00119-8.
5. Casadei, P., Nanni, A., & Alkhrdaji, T. (2005). Steel-reinforced polymer: An innovative and promising material for strengthening infrastructures. *Concrete Engineering International, 9*(1), 54–56. https://doi.org/10.1108/9781786350220-016.
6. Chong, K. P., & Garboczi, E. J. (2002). Smart and designer structural material systems. *Progress in Structural Engineering and Materials, 4*(4), 417–430. https://doi.org/10.1002/pse.132.
7. Han, S., Zhao, S., Lu, D., & Wang, D. (2023). Performance improvement of recycled concrete aggregates and their potential applications in infrastructure: A review. *Buildings, 13*(6), 1411. https://doi.org/10.3390/buildings13061411.
8. Hu, W., Shu, X., & Huang, B. (2019). Sustainability innovations in transportation infrastructure: An overview of the special volume on sustainable road paving. *Journal of Cleaner Production, 235*, 369–377. https://doi.org/10.1016/j.jclepro.2019.07.227.
9. Karthik, S., Rao, R. M., Awoyera, P., Akinwumi, I., Karthikeyan, T., Revathi, A., & Saravanan, S. (2018). Beneficiated pozzolans as cement replacement in bamboo-reinforced concrete: The intrinsic characteristics. *Innovative Infrastructure Solutions, 3*, 1–8. https://doi.org/10.1007/s40940-018-0070-4.
10. Kazemi, M., Wang, H., & Fini, E. (2022). Bio-based and nature inspired solutions: A step toward carbon-neutral economy. *Journal of Road Engineering, 2*(3), 221–242. https://doi.org/10.21273/JRE.2022.2.3.221.
11. Kurpayanidi, K. I., & Makhmudova, N. (2016). Current trends establishment of innovation infrastructure in the industrial sector of Uzbek economy. *European Journal of Natural History, 4*(2), 44–48. https://doi.org/10.3797/scipharm.2006.04.003.

12. Kurtis, K. E. (2015). Innovations in cement-based materials: Addressing sustainability in structural and infrastructure applications. *MRS Bulletin, 40*(12), 1102–1109. https://doi.org/10.1557/mrs.2015.259.
13. Leshko, B. J. (2006). Bridge design innovation: Materials and construction. In *Structures Congress 2006: Structural Engineering and Public Safety* (Vol. 2, pp. 1–10). https://doi.org/10.1061/40862(2006)5.
14. Makul, N. (2020). Advanced smart concrete—A review of current progress, benefits, and challenges. *Journal of Cleaner Production, 274*, 122899. https://doi.org/10.1016/j.jclepro.2020.122899.
15. Mistri, A., Bhattacharyya, S. K., Dhami, N., Mukherjee, A., & Barai, S. V. (2020). A review on different treatment methods for enhancing the properties of recycled aggregates for sustainable construction materials. *Construction and Building Materials, 233*, 117894. https://doi.org/10.1016/j.conbuildmat.2019.117894.
16. Nodehi, M., & Nodehi, S. E. (2022). Ultra high-performance concrete (UHPC): Reactive powder concrete, slurry infiltrated fiber concrete, and superabsorbent polymer concrete. *Innovative Infrastructure Solutions, 7*(1), 39. https://doi.org/10.1007/s40940-021-00233-w.
17. Song, S., Hou, Y., Guo, M., Wang, L., Tong, X., & Wu, J. (2017). An investigation on the aggregate-shape embedded piezoelectric sensor for civil infrastructure health monitoring. *Construction and Building Materials, 131*, 57–65. https://doi.org/10.1016/j.conbuildmat.2016.11.098.
18. Wasim, M., Ngo, T. D., & Law, D. (2021). A state-of-the-art review on the durability of geopolymer concrete for sustainable structures and infrastructure. *Construction and Building Materials, 291*, 123381. https://doi.org/10.1016/j.conbuildmat.2021.123381.
19. Zareie, S., Issa, A. S., Seethaler, R. J., & Zabihollah, A. (2020). Recent advances in the applications of shape memory alloys in civil infrastructures: A review. In *Structures* (Vol. 27, pp. 1535–1550). Elsevier. https://doi.org/10.1016/j.istruc.2020.09.027.

CHAPTER 3

REHABILITATION OF INFRASTRUCTURE: BY ADVANCED POLYMER COMPOSITE

LEARNING OBJECTIVES

At the end of this chapter, readers will be able to:

1. Understand the advantages of FRP composites over conventional materials in rehabilitation
2. Differentiate between repair, strengthening, and retrofit in structural rehabilitation.
3. Understand the variability and tailor ability of FRP composite properties.
4. Differentiate between fibers and matrices commonly used in FRP composites.
5. Comprehend the manufacturing processes for seismic retrofitting and strengthening.
6. Understand the properties and characteristics of FRP composites for infrastructure rehabilitation.
7. Know the durability and performance of FRP composites in civil engineering applications.
8. Explore the diverse applications of FRP composites in strengthening and repair of structures.

INTRODUCTORY EXAMPLE

PolyStruct Solutions, a leading provider of innovative solutions in infrastructure rehabilitation, is at the forefront of addressing the challenges posed by aging structures. With a focus on practicality and effectiveness, PolyStruct collaborates closely with municipalities and infrastructure stakeholders to develop tailored approaches to enhance the durability and lifespan of critical infrastructure components. In the field of infrastructure rehabilitation, bridges often require significant upgrades to ensure they meet modern standards and maintain their structural integrity over time. Stepping up to tackle these challenges is PolyStruct Solutions. Drawing upon extensive expertise in Fiber-Reinforced Polymer (FRP) composites, PolyStruct applies a systematic approach to rehabilitation planning.

Through rigorous testing and implementation, PolyStruct demonstrates the tangible benefits of incorporating FRP composites into bridge rehabilitation projects. By replacing conventional materials with lightweight yet durable alternatives, PolyStruct achieves notable improvements in structural performance and longevity. As PolyStruct continues to refine its methodologies and expand its project portfolio, the company remains steadfast in its commitment to driving positive impacts in infrastructure sustainability. By harnessing the capabilities of advanced polymer composites, PolyStruct contributes to the resilience and longevity of critical infrastructure networks.

This introductory example highlights PolyStruct Solutions' pivotal role in advancing infrastructure rehabilitation practices, offering pragmatic solutions that address real-world challenges and promote sustainable development.

UNIT INTRODUCTION

Across the globe, civil infrastructure facilities are a massive financial commitment and play a crucial role in providing comfort, security, and progress for humanity. After undergoing significant expansion over the past five decades, the infrastructure is now facing a critical age where evidence of deterioration and decreased efficiency are becoming more apparent. There are various issues with the present-day bridge inventory, such as corrosion, environmental damage, and aged structural components (Hollaway & Cadei, 2002). Additionally, the bridges are facing challenges due to higher traffic requirements and changing patterns. Some of these problems can be attributed to inadequate detailing during the initial design and construction, as well as the application of inadequate materials and construction techniques (Karbhari, 2004). Furthermore, factors such as irregular inspection and insufficient maintenance, along with updates in design codes, have necessitated significant rehabilitation and renewal. These shortcomings are not limited to only bridges and other structures related to transportation but are widespread throughout the constructed environment. Major cities are grappling with the unintended consequences of unchecked urban development, which has resulted in hasty construction using low-quality materials and inadequate attention to detail. This issue is particularly evident in the rapid urbanization of South-East Asia (Hulatt et al., 2003). Furthermore, apart from the deterioration caused by use and issues related to building codes, it is crucial to implement rehabilitation strategies to address the growing demands and evolving requirements of occupants and essential services. These factors greatly affect society, leading to significant socio-economic losses due to limited usable space, as well as delays and incidents or collapses (Balaguru & Kurtz, 1998).

Traditional materials like concrete, steel, and hardwood offer several benefits, including the affordability of their raw materials. Nevertheless, it is evident that traditional materials and technologies, while appropriate in certain situations and proven to be effective, fall short in terms of durability in some instances and are prone to rapid degradation in others. Highlighting the importance of improving the quality of these materials and exploring innovative technologies to enhance the range of options for civil designers and engineers (Chen & Hsiao, 1992). Utilizing FRP composites allow for the next level of advancement, as it enables the seamless integration of reinforcing fibers, along with suitable fillers and additives, within a polymeric resin matrix. It is important to note that the fibrous reinforcement plays a crucial role in carrying load in specific directions, as indicated by the fiber orientation. The polymer, on the other hand, serves as a medium for transferring stresses between adjacent fibers through adhesion. Additionally, it provides the material with toughness and protection (Hollaway, 2003). Composites made from FRP have become extremely popular in the field of construction materials due to their numerous advantages. These include exceptional specific stiffness and durability, lightweight properties, resistance to corrosion, and the potential for long-lasting durability. Additionally, FRP composites offer the capacity to tailor mechanical characteristics in the form of anisotropic materials. And also possesses low conductivity, acting as an effective insulator (particularly in the context of glass fiber strengthened composites), and exhibit electromagnetic transparency (Al-Salloum & Almusallam, 2003). Moreover, the extensive range of manufacturing processes allows for the fabrication of these materials in diverse settings, ranging from loosely controlled environments at the field (as seen in wet layup for external strengthening applications) in order to precise tolerances for dimensions in tightly regulated factory conditions (as observed in the autoclave cured combinations) (Helbling et al., 2006).

Composites made from FRP have found extensive use in various civil engineering applications, including bridge decks, structural profiles, tendons, strengthening bars, rehabilitation, and aesthetic fascia. However, this chapter will specifically concentrate on the application of these FRP composites within the rehabilitation of structural components (Friedrich & Almajid, 2013).

3.1. REHABILITATION AND FRP COMPOSITES

Learning Objectives

- Differentiate repair, strengthening, and retrofit in structural rehabilitation for clarity.
- Highlight advantages of FRP composites over steel plates in rehabilitation.

In this chapter, it is important to distinguish between the terms repair, strengthening, and retrofit, as they each refer to distinct structural conditions. Although these terms are frequently utilized interchangeably, falling under the broader category of rehabilitation (Figure 3.1). When it comes to addressing structural issues, the utilization of FRP composite materials is employed to rectify any deficiencies, such as cracks, and bring the component back to its initial level of functionality (Li et al., 2021). On the other hand, the enhancement of structures or elements is limited to situations where the inclusion of the FRP composite boosts the current performance level as intended, resulting in stated performance.

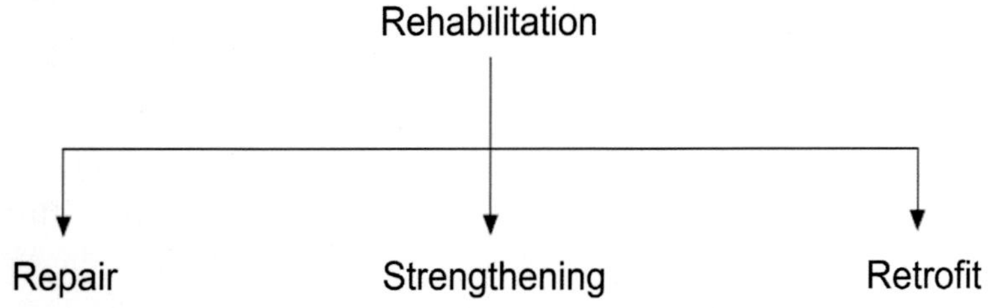

Figure 3.1. *Representation of the categories of rehabilitation.*

Source: JF Dawalos Creative Commons License.

This pertains to the reinforcement of bridge decks, where the utilization of external bonded composites made of FRP (Figure 3.2) can enhance the load-carrying ability (Seica & Packer, 2007). In this particular context, the term 'retrofit' is utilized to describe the seismic upgrade of facilities. An example of this is the solitary confinement of columns through the use of FRP jackets. It is crucial to differentiate not just for functionality, but also because the choice of substances and procedures should be tailored to meet the particular demands of the application, such as durability and anticipated service life (Pantelides et al., 2000).

Figure 3.2. *Illustration of application of composite strips to the soffit of a bridge deck.*

Source: CW Roeder Creative Commons License.

This chapter focuses on the application of composites made of FRP in rehabilitation projects. Specifically, examine cases where the FRP composite material is bonded to or in close interaction with the relevant section that makes up the concrete substrate. The techniques of seismic retrofitting are primarily based on previous research on jackets made of steel and concrete (Dong & Ansari, 2011). These jackets involve the addition of extra material to confine the concrete core, ensuring its stability. When it comes to strengthening and fix, the process is based on the use of steel plates which have been adhesively bonded or bolted on the substrate. This technique was developed for giving extra strength and stiffness. Despite its significant success and ongoing utilization, the methodology does have a few notable drawbacks,

- Steel plates can be quite heavy, and when added to a structure that is already weak, can potentially cause further damage or even lead to unintended failure (Duic et al., 2018).
- Precautions are required to ensure that the plates remain attached to the substrate due to their weight, often resulting in the need for time-consuming anchoring.
- Handling the plates in the field can be challenging due to their weight. It is necessary to utilize lifting equipment and to ensure that the plate is near the substrate throughout the adhesive curing process.

Additionally, positioning the plates in areas that have current utilities can be problematic (Lau et al., 2002).

- Using long continuous lengths can be difficult because of the weight or rigidity of the material.
- Field welding, particularly butt-welding for joining shorter lengths, may cause harm to the current framework, act as a focal point for future rusting, and give rise to concerns with non-destructive evaluation (Neale, 2000).
- There are concerns about the long-term durability, particularly in humid conditions.
- There are several concerns regarding the bonding that occurs during traffic flow as a result of sagging under its weight.
- The process can be quite involved and time-consuming due to the weight or difficulty in processing (Custódio & Cabral-Fonseca, 2013).
- When it comes to comparing different materials, FRP composites stand out for their lightweight nature, ease of application, impressive durability, and ability to be quickly and flexibly used in the field without causing major disruptions to traffic (Frigione, 2007).

3.2. MATERIALS AND MANUFACTURING PROCESSES

Learning Objectives

- Understand characteristics and fabrication of FRP composites for rehabilitation.
- Differentiate methods for seismic retrofitting and strengthening/repair using FRP composites.
- Know the advantages and challenges of various FRP composite applications in rehabilitation.

Although the field of civil engineering is familiar with the performance features of conventional materials like concrete, steel, and timber, there is still limited knowledge about the new category of FRP composites. In addition, there is a limited availability of standardized handbooks that encompass these properties (Rajak et al., 2019). This is primarily since the application of composites made from FRP in rehabilitation is a relatively new area of study and the features of these composites can vary (Wu et al., 2023). It is important to highlight that the characteristics of FRP composites are influenced by the characteristics of the material components, such as resin matrices, reinforcing fibers, and fillers, as well as their respective volume fractions and fiber orientation (Garden, 2004).

Therefore, FRP composites offer a range of customizable materials, allowing for properties to be adjusted to suit the specific requirements of the application (Jose & Joseph, 2012). This presents a paradox for designers, as it offers the freedom to break away from uniformity and limited options, but also poses challenges when it comes to routine design due to the absence of standardizing and the overwhelming number of choices available. Therefore, the reader must acquire a fundamental comprehension of the inherent properties of this emerging group of materials before delving into their potential applications in the rehabilitation of concrete substrates (Regassa et al., 2019).

3.2.1. Materials Overview

This chapter focus on FRP composites, which are materials created by combining durable and stiff fiber components with polymeric resins. These materials have unique characteristics that differ from their components, and it is possible to physically distinguish and separate all of them. In this discussion, the material systems and their constituents are primarily addressed that are commonly employed in rehabilitation, instead of focusing on potential future systems. Therefore, the discussion is limited to glass, carbon, and aramid fibers, as well as epoxy and vinylester resins for the matrices (Hsissou et al., 2021).

Table 3.1 presents a comparison between the fiber properties and grade 60 steel for construction. Glass fibers have a wide application as a common reinforcement type in composites, making up approximately 90% of all fibers utilized. They have been extensively employed in the field of rehabilitation (Abliz et al., 2013).

Table 3.1. *Properties of Typical Reinforcing Fibers*

Fiber Type	Designation	Density (G/ Cc)	Tensile Modulus (GPa)	Tensile Strength (MPa)	Ultimate Strain %)
Glass	E-glass S-glass	2.55 2.49	72.4 86.9	3450 4800	4.8 5.2

Carbon (PAN based)	T-300 AS4 IM6 UHM	1.76 1.80 1.74 1.85	230 221 275 441	3530 3930 5240 3445	1.5 1.7 1.7 0.8
Carbon (pitch based)	P-55 P-100 UHM	2.00 2.15 2.20	380 758 965	1900 2410 2410	0.5 0.32 0.27
Aramid	Kevlar 49 Kevlar 149 Technora	1.45 1.47 1.39	131 179 74	3620 3500 3500	2.8 2.0 4.6
Steel	Grade 60	7.80	200	Yield = 415 Ultimate = 620	4.5–12

Source: JF Dawalos Creative Commons License

The process of creating glass involves the fusion of silicates using potash, lime, or different metallic oxides. The production process for glass filaments involves the combination of silica and specific minerals within a furnace to create a homogenized melt. These filaments usually have a diameter ranging from 3-20 μm. After undergoing the refining process, the molten material is transformed into fibers using extrusion by means of platinum-rhodium bushings (Ghori et al., 2018).

The filaments cool down quickly to avoid crystallization. The filaments are extremely abrasive and can easily degrade when exposed to moisture. As a result, they undergo treatment with a sizing/binder shortly after they are formed and before they are gathered into strands. Various types of glass fibers are classified according to their chemical structure, each with its own designated letter code that indicates their unique properties and applications. E and S-Glass fibers are frequently utilized in structural applications, while rehabilitation uses are primarily limited to E-glass fibers. Their strength and modulus are commendable and can be easily shaped. However, it is prone to degradation when exposed to humidity and alkalis without the protection of a suitable resin matrix. Additionally, these materials are susceptible to creep and stress rupture, making them unsuitable for applications involving sustained loads of high level (Joshi & Chatterjee, 2016).

The origins of carbon fibers can be traced back to the experiments conducted by Thomas Alva Edison, who carbonated cotton and bamboo to create filaments for incandescent lamps. However, it was not until the research of Bacon and Tang that carbon fibers were developed for practical use in structural uses, using rayon or cellulose as a precursor (Guo et al., 2021). Union Carbide later explored the application of isotropic pitch as a precursor. Additionally, carbon fibers were also developed from polyacrylonitrile (PAN) by researchers Shindo and Watt. Later, Otani effectively created high-modulus fibers using pitch. Materials that serve as the foundation for carbon fibers are known as precursors. There are currently three main kinds being used: rayon, polyacrylonitrile (PAN), and pitch. The characteristics of the leading fiber, along with the economic aspects of the process, vary depending on the selection of the precursor material. Precursors made from cellulose, known as rayon, exhibit a relatively low conversion efficiency (Ahmad & Ahmad, 2009).

Aramid fibers are a type of organic fiber made from aromatic polyamides. They are typically produced by extruding a polymer solution using a spinneret. The expanded chain structure found in these liquid crystalline polymers incorporates aromatic rings, resulting in exceptional thermal stability. Additionally, the presence of amide (-NH-) and carbonyl (-CO-) bonds within the structure enhances resistance to rotation and imparts superior strength and modulus. The microscopic makeup of aramid fibers exhibits a fibrillar pattern, resulting from the arrangement of prolonged, parallel polymer chains. This unique structure imparts an anisotropic characteristic to the fibers, with greater strength and modulus observed in the axial direction rather than the radial (transverse) path (Aisyah et al., 2021). Due to the unique fibrillar framework, the fibers are prone to the development of kink bands or microscopic buckles when subjected to compression. The behavior of fibers under tension is linear, while under compression it becomes plastic, exhibiting yield at compressive strain levels as low as 0.3–0.5%. There are several types of aramid materials available, and it is important to keep in mind that Kevlar is a particular kind of Poly-paraphenylene Terephthalimide, often mistakenly utilized as a generic term for aramids (Nayana & Kandasubramanian, 2020).

Typically, the fiber component of a composite made from FRP undergoes some modifications before it is utilized in the manufacturing process. Instead, a group of filaments (such as bundles, tows, yarns, etc.) is utilized in a fibrous assembly or a fabric form. The one-way design serves as the main foundation for a collection of directed, and non-woven, materials. During this process, strands of roving, yarn, or tow are arranged in a parallel manner and secured together with threads that are positioned perpendicular to them. These threads can either stitch the components together or be bonded to them through a heating method.

Woven fabrics are utilized in circumstances and are created by interconnected multiple sets of bundles in a prescribed orientation. The most prevalent kinds of this type are biaxial weaves, in which each pair of bundles have a perpendicular orientation to one another, and triaxial weaves. While plain weaves are the most prevalent type of assembly, there are additionally other yarn architectures that involve perpendicular yarns interwoven at certain intervals. These alternative architectures are employed to achieve characteristic ratios and to facilitate conformance. These assemblies are commonly referred to as 'satins' (Oladele et al., 2020).

3.2.2. Manufacturing Processes

For this discussion, the term 'manufacturing processes' is discussed to refer to the specific processes involved in rehabilitating concrete utilizing externally bonded composites made of FRP. For a more focused approach, it is beneficial to categorize this into two main areas (Frketic et al., 2017): (i) techniques employed for enhancing the seismic resistance of columns and (ii) techniques utilized for improving the strength and repairing beams, girders, slabs, and walls. In the upcoming sections, the discussion on manufacturing in FRP composites is done (Goh et al., 2019). However, it is important to note that this discussion cannot be separated from the materials and configuration aspects. Therefore, these aspects will also be taken in consideration, highlighting the interconnected nature of these three considerations (Parmar et al., 2022).

3.2.2.1. Seismic Retrofitting

Concrete columns that require retrofitting often lack the necessary bar buckling restraint, flexural ductility, shear strength, or lap splice clamping. Applying well-designed jackets/wraps surrounding the columns creates lateral confining stresses within the concrete. These stresses occur when the concrete expands laterally due to high axial compressed strains or in the tension zone due to the dilation of lap splices (Iacobucci et al., 2003). Which can also result from the formation of diagonal shear cracks. Rehabilitation of infrastructure often involves implementing conventional retrofitting procedures for reinforced concrete columns. These measures typically include externally confining the core with heavily reinforced concrete sections, using steel cables wound helically surrounding the existing column with close spacing and covering them with concrete, and employing steel shells or casings or which are welded tightly together in the field to confine the present columns (Rajput et al., 2019).

While certain methods have proven to be highly efficient but require a considerable amount of time for installation. Additionally, that may result in significant traffic disruptions as they necessitate access and space for heavy equipment. Another concern is the reliance on field welding, which can be unreliable in terms of quality and consistency. Furthermore, these methods are prone to degradation caused by corrosion. In addition, the material used in steel casings or jackets has an isotropic nature, which means that the jacket not only provides the necessary confinement, but also enhances stiffness and strength capacity for the retrofitted column. These attributes are considered undesirable as they often result in the transmission of greater seismic force levels to neighboring structural elements (Fahmy, 2023).

Fiber-reinforced composites offer a solution for solitary confinement without the need for additional stiffness (using only hoop reinforcement, with no axial reinforcement). May allow for the quick production of affordable and long-lasting jackets, minimizing traffic disruption in many situations. In a general sense, there are six main categories for categorizing fiber-reinforced composite covers (or jackets), based on the process and/or materials form. This is illustrated in Figure 3.3 (Honarparast, et al., 2019).

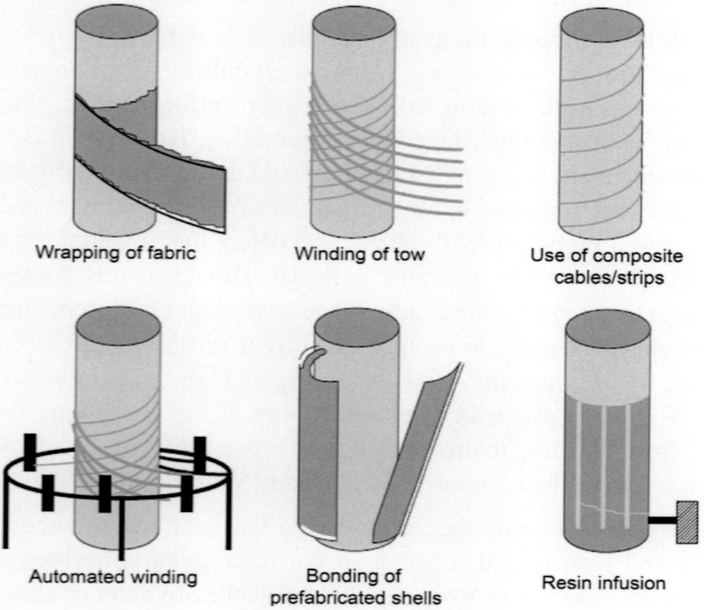

Figure 3.3. *Methods of application of FRP composites for seismic retrofit and strengthening of columns.*

Source: P Qiao, Creative Commons License.

During the wet lay-up procedure, the fabric is saturated on site and then applied to the column, allowing it to cure naturally under normal conditions.

In damp winding, the manufacturing procedure is automated but still follows a similar pattern. However, the resulting jacket exhibits nominal pre-stress due to the application of winding tension. The damp layup process is typically linked to manual usage and the use of ambient cure (Arockiasamy & Dutta, 1996). However, it is also possible to apply heat to the entire system after application to achieve higher cure temperatures and, consequently, a higher glass transition temperature (Tg). The process offers a great deal of flexibility, particularly in confined areas. However, there are some concerns regarding quality control for the resin mix, ensuring that the fibers are properly saturated with resin without trapping too many air pockets, achieving proper compaction of the fibers without leading to excessive wrinkling, managing the curing process effectively, and ensuring long-term durability in various environmental conditions. When it comes to the damp winding for tow or tape, there is potential for automation in the process. However, resin impregnation still relies on a wet shower or spraying, and a lot of the concerns remain like those discussed in the wet layup process (Pampanin et al., 2007).

When utilizing prepreg material, a higher cure is typically involved. The winding procedure for tow and tape is automated, while the fabric process requires manual lay-down. Both options utilize a higher curing method, which helps alleviate some concerns regarding Tg and durability. However, it is important to follow proper manufacturing procedures and avoid using anhydride-based systems, as they are sensitive to moisture during the curing process (Zou et al., 2018). Utilizing a prepreg-based mechanism offers the potential for enhanced control of quality over the incoming material. However, this comes with an additional cost and requires strict adherence to a proper cure schedule. Failing to do so would negate any advantages gained from the material form (Silva et al., 2007). In addition, the need for curing at high temperatures (typically between 175-300 °F) can lead to complications if the underlying concrete is excessively damp, leading to the formation of blisters due to water vapor during the curing process of the composite FRP jacket (Ma & Xiao, 1999).

When it comes to prefabricated shells, the sections are manufactured in a factory and then joined together in the field using adhesive bonding to create the jacket. There are two possible approaches to achieve this: utilizing two separate sections and a single section that is extended to encompass the column. The controlled factory-based production of the shells ensures a high level of arriving product quality control in this procedure (Vaghani et al., 2014). Nevertheless, the effectiveness and longevity of the system rely on the adhesive's capacity to bear weight, which in turn depends on the strength of the bond formed in the field and the adhesive's capacity to withstand a challenging and unpredictable environment. This environment may involve high levels of moisture (or even being submerged in water, such as in a flood plain) and significant temperature fluctuations. In most cases, the adhesives commonly employed in these applications are cured at room temperature and belong to the categories of polyurethanes or epoxies (Aref & Jung, 2003). It is important to mention that polyurethanes typically include components that are sensitive to moisture and may revert when exposed to heat and/or moisture. Despite their favorable characteristics in colder conditions, these materials often experience notable decreases in shear performance when exposed to higher temperatures. Epoxies, on the other hand, offer a wider range and exhibit minimal creep and shrinkage, albeit with a slightly lower peel strength overall. Regrettably, there has been a lack of focus on the detrimental impact of moisture and temperature on bond performance (Triantafillou, 2001). It is important to highlight that when the adhesive deteriorates, either from absorbing too much moisture or from both the effects of heat cycling and absorption of moisture, it can lead to the separation of layers and ultimately the failure of the entire system. This can occur even if the composite made from FRP shells themselves have not experienced significant degradation (Seyhan et al., 2015).

When it comes to resin infusion, the dried fabric gets applied by hand and the resin is infused into it using vacuum, with the curing process taking place under ambient conditions. Thus far, there has been limited exploration of the method that utilizes cables or prefabricated strips (Dai et al., 2012).

3.2.2.2. Strengthening/Repair

The deterioration caused by rusting of steel reinforcement, concrete cover spalling, significant concrete cracking caused by carbonation or freeze-thaw action, the impact of alkali-silica reaction; however, and the ever-changing demands of traffic (both in terms of vehicle volume and load capacity) have resulted in an urgent requirement for effective techniques for repairing and strengthen beams, girders, deck soffits, and bridge deck slabs. There is a strong push to improve the load capacity of bridges in various regions, like Europe, Australia, and Japan (Marshall et al., 1996). This is driven by the need for uniform posters and the safe transportation of heavier truck axle weights. To meet these demands, it is crucial to adopt technology that can effectively and economically achieve these goals without causing major disruptions to traffic. Over time, buildings may experience wear and tear, as well as changes in their purpose and requirements. This often calls for improvements to the infrastructure, such as reinforcing existing slabs and beams. For example, in older buildings, floor slabs may need to be cut to make way for elevator shafts during renovations. The effectiveness of this method relies on the interaction between the composite element and the concrete substrate, specifically focusing on the strength of the bond between it and the interface layers (Truong et al., 2017).

Therefore, it is crucial to view the composite-adhesive/resin-concrete system as a comprehensive entity. The properties of each component, as well as their interactions with one another and the surrounding environment, can greatly impact the overall effectiveness, longevity, and resilience of the system. There are three different methods for fabricating the 'plate' (of external reinforcement), which include resin infusion, adhesive bonding of prefabricated elements or wet layup of fabric. Among these options, the pre-manufactured alternative stands out for its exceptional uniformity and quality control. This alternative involves fabricating the strip or plate by means of FRP composite procedures like pultrusion, pulforming, or wet layup utilizing a vacuum bag, all of which are carried out in controlled factory conditions (Brindha et al., 2023).

Figure 3.4. *Description of close-up showing splitting within the prefabricated strip.*

Source: CW Roeder Creative Commons License.

The adhesive bonding method involves the prefabrication of the composite strip/panel/plate, which is then securely attached to the concrete substrates via an adhesive and pressure. The speed of application is crucial, and the effectiveness of the process depends on selecting the right adhesive and ensuring a strong bond between the substrate of concrete and the composite's adherend. It is crucial to select an adhesive that closely matches the elastic moduli along with coefficients of expansion due to heat of both concrete itself and composite materials (Buyukozturk et al., 2004). This will help minimize the stresses caused by any mismatch, with the added benefit of an interlayer to further reduce these stresses. Environmental effects are further addressed in subsequent sections. Strips and plates may be pre-produced using different fabric forms. However, the readily accessible strips are currently made utilizing unidirectional carbon fiber reinforcement. This reinforcement is subjected to specific thicknesses and widths. Using a unidirectional form maximizes the theoretical effectiveness of the fiber. However, it requires thicker materials to prevent longitudinal splitting and can experience interlaminar cracking among filament layers. In Figure 3.4, there is an example that illustrates the movement of the failure surface from within the concrete within the composite strip (Sen, 2016).

The alteration in the failure path can be attributed to the utilization of a pultruded unilateral carbon/epoxy strip to serve as the external reinforcement. The composite employed in this case has a fiber volume fraction of approximately 70%. This results in a comparatively small resin content, leading to the formation of microscopic dry areas and/or voids among individual filaments. These weak zones can easily propagate fractures. In addition, the resin utilized was quite fragile, which means that when a crack reaches the interface between the adhesive and the composite, it is more likely to propagate through this area. This typically occurs when there is a flexural crack within the concrete that extends into the adhesive (Rubino et al., 2020).

In the wet layup procedure, resin is applied to a concrete substrate and then layers of fabric are impregnated and bonded on the substrate via the resin. The composite along with the bond are created simultaneously in the field. The process offers great flexibility within the field, but it does have the drawback of field mixing along with fabrication. There is a significant concern regarding the proper mixing of a resin system as well as the potential for moisture absorption or the introduction of impurities. Furthermore, the process carries the inherent risk of uneven saturation of the fabric and the formation of wrinkles, both of which can impact the final properties (Mosallam, 2004). The wet layup procedure is widely utilized in various applications due to its high flexibility for field use and cost-effectiveness. Nevertheless, it exhibits significant variability and requires the application of excessive resin, potentially leading to the wrinkling or deformation from shear of the fabric employed, thereby reducing its intended strengthening effectiveness. In addition, the process is accompanied by the inherent risk of air voids being trapped, which can lead to deterioration over time. The composite created in this scenario is typically cured under normal environmental conditions. Currently, there are commercially available options for both simple weave and unidirectional fabrics. In Japan, they have even developed kinds of unidirectional fabrics using paper backing to make rehabilitation in the field easier, resembling the application of wallpaper (Bakis et al., 2002).

The in-situ resin infusion technique is a relatively recent variant that allows for uniformity and effective fabric compaction. It also simplifies the placement of reinforcement, minimizing unintended deformation. Implementing this scheme on extensive areas poses challenges and requires maintaining vacuum pressure, which can be particularly challenging on severely cracked surfaces. During the resin infusion procedure, the reinforcement fabric is initially shaped into a preform and then securely bonded to the substrate via a vacuum bag. The composites are formed by infusing resin into a fibrous assembly under vacuum. In the method known as wet layup procedure, the composite as well as bond are formed simultaneously. As part of an alternative approach, the perform assembly undergoes a treatment where the outer layer is infused and partially solidified beforehand, ensuring a high-quality outer surface. During the process, there is a possibility of uneven filling or incomplete filling of the preform. This can occur because of local irregularities, surface conditions, and the inability to maintain full vacuum (Pöhler et al., 2021).

As previously mentioned, the effectiveness of an external composite relies heavily on the connection between the composite and the concrete. While extensive research has been carried out in other fields of bonding, particularly about metal bonding, there has been a lack of focus on surface analysis and treatment when it comes to strengthening concrete. Unfortunately, the attention given to this aspect has been limited to sandblasting and abrasion. There are several potential options and connections available for the three scenarios of wet layup, adhesive bonding, and resin infusion. Typically, any small imperfections or fractures in the concrete are addressed by applying a putty, paste, or filler. This layer is expected to have a more gentle and compliant nature when it comes to the propagation of cracks and fractures. When resin infusion is used, the infusing resin can behave like a crack injection, effectively filling any cracks and causing notable changes to the substrate material (Cevik & Cabalar, 2008). It is important to consider that when surface flaws are deep or have a non-uniform shape/configuration, there may be instances where they are not filled when applying an adhesive or primer layer on the concrete's outermost layer. This can potentially result in air voids, gaps, or entrapment hollows that may significantly affect the durability and fracture conduct. Within the wet layup and resin infusion approaches mentioned earlier, the resin itself serves as the adhesive, creating a bond using the concrete substrate during the fabrication of the composite. There are pros and cons to this situation. On the positive side, by removing the third phase within the system, which is the adhesive, there are fewer interfaces where failure could happen. Additionally, the need for a more flexible layer is also eliminated. The occurrence of cracks at these points and/or the alterations in properties of the adherents over time and in different environments can lead to a rapid breakdown of these rehabilitation strategies (Uktamovich et al., 2016).

Insufficient surface preparation may result in defects in the bond line, which can compromise the effectiveness of the strengthening process and potentially cause the system to fail prematurely. Addressing this issue can prove to be challenging and require a significant investment of time, especially in certain situations. To address this issue, the application of nearby surface mounted (the NSM) FRP offers a solution to overcome this challenge in certain cases. This involves placing the fiberglass reinforced plastic composite reinforcement within a groove that is cut into the concrete. This approach represents an update of the traditional rebar slotting method, as discussed by Johnson in 1965. When it comes to rebar, a groove is carefully carved to the correct depth and width. Then, the rebar is placed into the groove and securely fixed using a strong adhesive. Switching to FRP reinforcement offers several benefits, including resistance to corrosion, a lightweight structure, customizable thermal properties, and efficient use of the reinforcement (Khalid Heiza et al., 2014).

Due to the non-corrosive nature of the FRP, the depth of the groove is reduced as the need for cover is eliminated. Furthermore, the fiberglass reinforced plastic composite shape may be fine-tuned to maximize surface area to shear transfer, like flat reinforcement. This not only reduces interfacial stress but additionally enhances stress transfer among the substrate or the FRP, boosting overall effectiveness (Moy, 2013).

3.3. CHARACTERISTICS AND PROPERTIES OF FRP COMPOSITES

Learning Objectives

- Understand typical properties of FRP composites for civil infrastructure rehabilitation.
- Understand the durability challenges and degradation mechanisms in FRP composite materials.
- Know the reliability-based approaches for predicting service life of rehabilitated structures

The properties of FRP composites are determined by the materials utilized, the way they are arranged (including fabric architecture, layup, and orientation), and the technique of processing. Typically, the properties of fiberglass reinforced plastic composites utilized for strengthening are like those of the standard processes employed. Tables 3.2–3.4 present the characteristic properties of fabrics, resins, and the end-product composites through the wet layup process (Han et al., 2023). The characteristics of prefabricated strips can be found in Table 3.5. As anticipated, the prefabricated strips exhibit considerably greater tensile strength and modulus compared to the composites produced through wet layup processing. This is attributed to the ability to achieve higher fiber volume proportions in these materials. It is important to consider that materials based on wet layup exhibit a notable degree of variability. This variability can impact the consistency of rehabilitation within the field, the determination of design allowable over the material at hand, and resultant factors of safety, which may differ from the assumptions made during the design process (Berardi & Dembsey, 2015).

Table 3.2. *Common Attributes of Fabrics Commonly Employed in Infrastructure Rehabilitation*

Fabric Type	**Style (Warp/Weft)**	**Aerial Weight (g/)**	**Weight %**	**Heat Set**	**Theoretical Thickness (mm)**
Glass Fabric					
100	Weave (E-glass/polymide)	920	98%/2%	Yes	0.36
107	Weave (E-glass/aramid)	920	98%/2%	Yes	0.36
Carbon Fabric					
103	Weave (24 k carbon/polyamide)	610	99%/1%	Heat set	0.34
117	Weave (12 k carbon/polyamide)	300	99%/1%	Heat set	0.17
230	Weave (12 k carbon/polyamide)	230	99%/1%	Heat set	0.12

Source: VM Karbhari Creative Commons License

Given the limited familiarity of FRP composites among civil engineers and infrastructure planners, there are significant concerns regarding the long-term durability of these substances, particularly over a span of 18 months (Cao & Wu, 2008). By comparing Figures 3.5 and 3.6, it becomes evident that the system experiences a greater decrease in tensile strength during each exposure period. Interestingly, the final change within the transition temperature of the glass is approximately equal. This suggests that the system has a similar ability to maintain its performance under challenging and fluctuating environmental conditions while under load (Toutanji & Gomez, 1997).

Table 3.3. *Properties of Common Resins Utilized in Infrastructure Restoration Common Attributes of Fabrics Commonly Utilized in Infrastructure Rehabilitation*

Characteristic/property	HEX-3R Epoxy 300	HEX-3R Epoxy 306
Service temperature range (C)	–40 C to +60°	-40°C to +60°
Application temperature range C	10°C to 40°	10°C to 40°
Shelf life (years)	2 years (unopened)	2 years (unopened)
Pot life (hours)	4 hours at 21°C	3 hours at 21°C
Cure time (days)	5 days at 21°C	5 days at 21°C
Gel time (hours)	14 hours at 21°C	5 hours at 21°C
Tack free time (hours)	20 hours at 21°C	16 hours at 21°C
Viscosity (cps)	Part A: 13000 cps at 25°C Part B: 17 cps at 25°C	Part A: 44000 cps at 25°C Part B: 18 cps at 25°C
Glass transition temperature (°C)	51°C after 7 days at 23°C 79°C after 5 days at 23°C and 48 hours at 60°C	48°C after 7 days at 25°C 80°C after 5 days at 25°C and 48 hours at 60°C
Tensile strength (MPa)	68	65
Tensile modulus (GPa)	3.4	3.56
Elongation at failure (%)	3.1	3.5
Flexural strength (MPa)	116	103

Flexural modulus (GPa)	3.19	2.92
Cured specific gravity	1.16	1.14

Source: L Shan Creative Commons License

While FRP composites have proven effective in various industries such as industrial, automotive, marine, and aerospace, it is important to note that there are significant distinctions in terms of loading, environment, and the materials and processes utilized in these applications compared to those that are likely to be employed in civil infrastructure. There is strong evidence to suggest that if these systems are designed and fabricated properly, they can have longer lifetimes and less upkeep compared to structures made from traditional materials. Unfortunately, there is a lack of concrete data on durability that is both well-documented and readily accessible to civil engineers (Maheri & Adams, 1999).

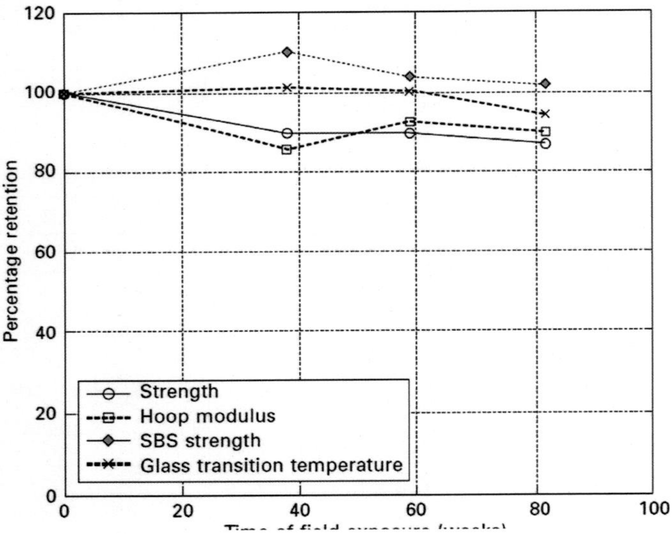

Figure 3.5. *An analysis was conducted to determine the relationship between the retention of properties and the duration of exposure for an adhesively bound E-glass/polyester prefabricated shell system to understand how the features change over time.*

Source: HJ Dagher, Creative Commons License.

Ongoing research is being conducted to evaluate the durability and lifespan of these materials. For instance, a study examined the changes in features of two column wrap processes that were subjected to field circumstances in Seattle, Washington for a period of 18 months. By comparing Figures 3.5 and 3.6, it becomes evident that the system experiences a more significant decrease in tensile strength during each exposure period. However, the ultimate change in the temperature of the glass transition remains relatively consistent (Amran et al., 2018).

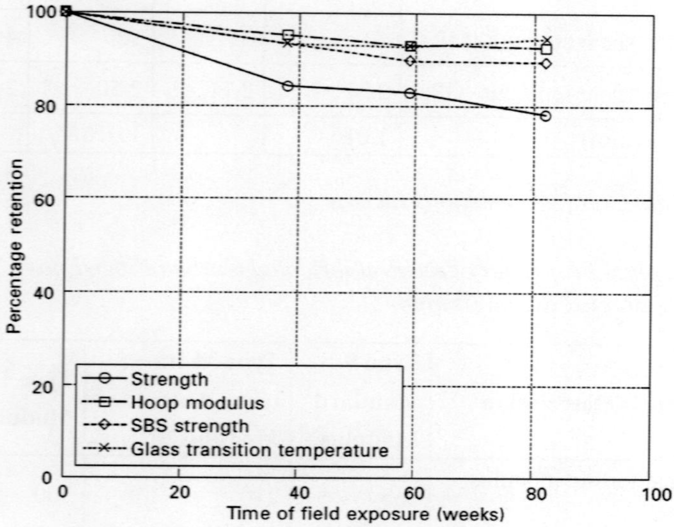

Figure 3.6. *Description of the Percentage retention of properties as a function of exposure time for a wet layup based E-glass/epoxy column wrap system.*

Source: HJ Dagher Creative Commons License.

Table 3.4. Characteristics of Laminates Frequently Utilized in FRP Composites Created via the Wet Layup Method for the Restoration of Infrastructure

Property	**Type 100 E-Glass Fabric**		**Type 103 Carbon Fabric**	
	Average	**Design**	**Average**	**Design**
Tensile strength (MPa)	612	558	849	717
Tensile modulus (GPa)	26.12	24.44	70.55	65.09
% Elongation	2.45	2.23	1.12	0.98
Tensile strength @ 140°F (MPa)	551	531	847	699
Tensile modulus @ 140°F (GPa)	25.69	23.36	69.84	63.09
% Elongation @ 140ºF	2.28	2.14	1.13	0.97
Compressive strength (MPa)	597	542	779	715
Compressive modulus (GPa)	29.72	25.49	67.01	61.53

90° Tensile strength (MPa)	30	30	24	16
90° Tensile modulus (GPa)	6.65	6.32	4.86	3.97
90° Elongation (ultimate percent strain)	0.46	0.34	0.45	0.33
±45° In-plane shear strength (MPa)	40	34	52	46
±45° In-plane shear modulus (GPa)	2.31	2.11	2.50	2.39
Ply thickness (mm)	1.016		1.016	

Source: P Qiao Creative Commons License

Table 3.5. *Typical Properties of the Prefabricated Carbon Fiber/Epoxy Strips used in Adhesively Bonded Rehabilitation*

Characteristic Description	**Type S Standard Modulus**	**Type M Intermediate Modulus**	**Type H High Modulus**
Longitudinal tensile modulus (GPa)	165	210	300
Longitudinal tensile strength (GPa)	2.8	2.4	1.3
Ultimate strain (%)	>1.7%	>1.2%	>0.45%
Density (g/c)	1.5	1.6	1.6

Source: (Walter M, Creative Commons License)

It is important to acknowledge that the absorption of moisture can lead to a decrease in the temperature at which glass transitions and alterations in mechanical properties. During the second phase of exposure, there is a noticeable increase in the mean difference of results in both systems. This increase can be attributed to variations in moisture absorption levels or gradients throughout the thickness of the material. After an extended period of exposure, the moisture level becomes more consistent throughout the entire thickness, leading to a reduction in scattered boundaries (Bank, 1993).

The strength of the composite is primarily determined by its intrinsic properties. However, in the context of seismic retrofit, the inter-layer effects have a significant impact on the hoop modulus. The effectiveness of the manufactured adhesively bonded shell system relies on the adhesive to transfer stress between the various layers for the factory-fabricated shell segments. However, when the adhesive properties change due to exposure, it leads to increased flexibility and inter-layer slip. Consequently, there is a significant decrease in the hoop modulus over time (Jagadeesh et al., 2023).

The flexibilization also leads to a higher perceived toughness, resulting in a minimal alteration in inter-laminar shear properties as determined by short-beam-shear testing. It should be acknowledged that this indicates a decline in adhesion and a likelihood of further degradation in performance as a whole, such as the initiation of disbonding, particularly if the adhesive absorbs more moisture over time (Zaman et al., 2013).

Although safety factors, or limited safety elements, are utilized in the approaches recommended by ACI-440 (2002) or TR-55 (2000) to determine acceptable performance levels for FRP composites utilized for rehabilitation, it is important to note that these factors rely on predetermined values rather than estimates of reliability. Therefore, when these factors are taken into account along with predictive methods for estimating the duration of "safe" service life and can create a misleading sense of assurance. This could lead the designer to mistakenly believe that the likelihood of failure is close to zero until the value of the threshold is reached (Günaslan et al., 2014).

For instance, Figure 3.7 illustrates the variations in the likelihood of failure for a two-layer carbon/epoxy wet layup composite made of FRP when it is submerged in deionized water at a temperature of 23°C and subjected to flexure testing. The likelihood of failure at a particular flexural strength value rises over time of exposure. In this context, it is important to evaluate the design final strength threshold value specified in ACI-440 (2002), which is 379.64 MPa. Based on a safety perspective, it would take approximately 22.62 years to reach this threshold. Therefore, it is reasonable to assume that the composite can be used safely at levels below or equal to the threshold until that point in time (Agarwal et al., 2018).

It should be noted that the likelihood of failure reaches 0.21 within the first five years and a value of 0.48 within 15 years at this threshold. By considering statistical variation among material characteristics, a more accurate assessment of safety can be made. This enables the designer to estimate risk more effectively and select design covers that meet reliability demands. It is crucial to continue developing methods that can accurately predict the service-life of both the FRP composite utilized during the rehabilitation and the rehabilitated framework itself (Yu et al., 2007).

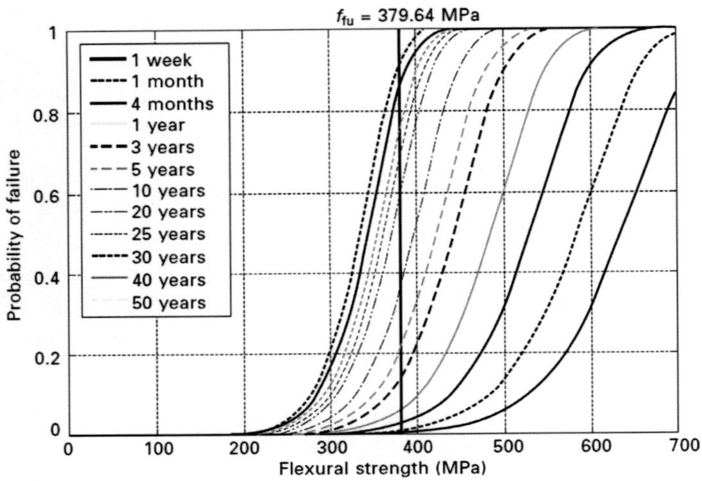

Figure 3.7. *Schematic Estimation of probability of failure for flexural strength response of two layered carbon/epoxy specimens immersed in deionized water as a function of time.*

Source: Harrison J, Creative Commons License.

3.4. APPLICATIONS OF FRP COMPOSITES

Learning Objectives

- Understand applications and advantages of FRP composites in concrete rehabilitation.
- Understand methods for strengthening and repairing concrete structures using FRP.
- Comprehend the effectiveness and cost benefits of FRP rehabilitation techniques.

Over the years, numerous restoration efforts have been successfully effectively carried out by utilizing FRP composites bonded in concrete substrates. These projects have been the result of extensive research and practical demonstrations. Within the field of seismic retrofit, various methods are employed, such as the wet layup process for retrofitting individual column bents, to the application of adhesively bonded shells for thousands of columns at the Yolo Causeway at Sacramento, CA (Xiao, 2004).

Externally bonded composites made of FRP have been widely used for strengthening purposes, particularly in the flexural support of beams and girders. This technique involves bonding FRP reinforcement onto the lowermost layer of a beam/girder or within the web area to enhance shear strength. A carbon-fiber bolstering system was utilized to repair and strengthen the shear capacity of more than 5000 precast pre-stressed double-tee beam stems in the Pittsburgh airport's parking complex.

This involved the use of stabilized unidirectional fabric, which was impregnated within the field utilizing the wet layup process. The project consumed a significant amount of material, totaling over 34,000 square meters. The project was successfully finished in less than four months, causing minimal closures and disruptions in the use of the garage.

This showcased the notable benefits of using FRP rehabilitation systems compared to traditional methods. At LaGuardia airport in 1998, a fascinating application involved the use of over 3000 m for pultruded carbon/epoxy strips to reinforce runway slabs in Flushing Bay in Queens, NY. The strips were utilized to enhance the overall strength ability of the slabs and reinforce them to avoid surface cracking and potential failure caused by the holes created for the insertion of additional support piling (Das & Nizam, 2014).

There is a growing recognition of the crucial role that slab strengthening or repair plays in ensuring the longevity and functionality of structures. It goes beyond simply increasing load capacity but also addresses the need to fix deficiencies in structures, like local punching shear failures observed in bridge decks among longitudinal girders. A recent case in point is the rehabilitation of a 103 m long, five-span bridge located over the California aqueduct in Northern California. The bridge, constructed in 1964 to be a cast-in-place concrete-reinforced T-girder bridge, has experienced significant deterioration over time.

This deterioration is a result of various factors, including human error and natural wear and tear. As a result, the deck of the bridge has developed longitudinal and transverse cracks, leading to damage caused by punching shear. Additionally, the overall loss in stiffness has caused the girders to crack. Considering the importance of maintaining the water flow in the aqueduct, preventing any contamination from construction materials, and minimizing traffic disruptions, it was determined that traditional repair methods were not suitable. After conducting an in-depth examination of the bridge's system, a rehabilitation plan was developed. This plan involved using composite materials to strengthen the slab within the two transverse and longitudinal directions. To ensure a clean and efficient installation process, a set of scaffolds alongside sheeting in between was utilized, preventing any debris from entering the water (Van Den Einde et al., 2003).

The study utilized carbon/epoxy pultruded strips and stitch-bonded linear carbon fabric impregnated within the field via an epoxy and wet lay-up. The pultruded strips were 51 mm wide with a total thickness of 1.2 mm and had a longitudinal modulus and a strength of 150 GPa and 3000 MPa, respectively. On the other hand, the stitch-bonded carbon fabric was 152 mm wide alongside an overall thickness of 1.1 mm achieved through multiple fabric layers. It had a longitudinal modulus along with strength of 75 GPa and 930 MPa, respectively. By utilizing composites, the rehabilitation was successfully carried out without the need to remove utilities such as gas pipelines that were hanging from the deck. This approach could not have been feasible if traditional methods like adhesively attached or bolted steel beams had been employed for strengthening. Modal tests conducted before and after the rehabilitation process confirmed the effectiveness of the strengthening scheme.

The results demonstrated that the deck slabs, which were in the worst condition, experienced a significant increase in stiffness, ranging from 25-35% (Nanni, 2001). This improvement more than made up for the prior deterioration. The utilization of composites allowed for a swift and seamless restoration, ensuring uninterrupted water and traffic flow. Moreover, this approach eliminated the necessity of removing and reinstalling utilities. Another example involves the restoration of a T-girder bridge constructed in 1969. This bridge, which spans 225 meters and consists of two lanes, is crucial for interstate traffic between California and Nevada. It is particularly important for accommodating permit trucks that surpass the standard load capacity.

The bridge exhibited notable indications of decay because of deck cracking and localized punching shear. Given the significant amount of traffic and the strategic location, it was necessary to prolong its lifespan to facilitate the completion of plans and secure funding for the buildup of a new bridge. Ensuring that traffic flow remained uninterrupted was the primary concern throughout the rehabilitation process. Due to various limitations, such as cost, an externally bonded composite choice was once again chosen as one of the most efficient solutions. The rehabilitation process involved the utilization of prefabricated pieces and fabric in various bays, having the spacing and the quantity of material customized to meet the specific needs of each bay (Gudonis et al., 2013).

The rehabilitation project was carried out seamlessly, ensuring uninterrupted traffic flow. It was completed in a remote area, where facilities along with heavy equipment were not readily available. The operation relied on the use of manually placed scaffolds and specialized equipment for the fabrication or bonding of composites. These tools were compact enough to be transported in a small pickup truck.

After the rehabilitation work was completed, the bridge was able to continue serving without any interruptions to traffic for a prolonged period. The initial assessment of the lifespan, as demonstrated by time-dependent reliability analysis, highlights the efficacy of this approach. The early research on the effectiveness of this technique suggests that the costs associated with rehabilitating infrastructure can be significantly lower compared to building a new bridge. The actual cost savings can vary depending on the extent of rehabilitation needed. Additionally, this approach can also extend the service life of the infrastructure by at least 10-15 years (Bedon, 2016).

Sectional openings, also known as cut-outs, are frequently required in existing slabs of buildings and factories due to evolving structural and functional needs. One possible reason for this is the requirement to install various elements like escalators, elevators, staircases, utility ducts (such as those for heating and air conditioning).

This often involves creating a cut-out in the infrastructure, which involves removing either the concrete or reinforcing steel. This may lead to a significant shortfall that needs to be addressed. Traditional approaches to mitigation, like incorporating edge girders or columns at cut-out corners or constructing load-bearing walls within the edges, can occupy precious floor space and may not meet aesthetic requirements. There are similar concerns related to the utilization of trusses for load redistribution.

While steel plates have proven effective for flexural strengthening, there are certain concerns regarding rust at the steel/adhesive connection as well as joints among plates. Additionally, the handling of massive plates in places with restricted access or limited clearance can pose challenges. Furthermore, incorporating heavy plates onto the tension-sensitive surfaces of slabs that have been weakened due to aging or deterioration might not have been considered favorable. Research has demonstrated that the application of strips of FRP within the opening is an effective method for restoring capacity and redistributing stress caused by the introduction of a cut-out in slabs of concrete. Furthermore, the incorporation of FRP strips not only enhances the reinforcement of the slab around the cut-outs, but also effectively decreases strain the amount in the steel, leading to the formation of a more favorable cracking pattern (Uomoto et al., 2002).

Practice Problem

Evaluate and compare three types of fibers commonly used in Fiber-Reinforced Polymer (FRP) composites: glass, carbon, and aramid fibers. Describe their respective properties and provide an example application for each type of fiber within the context of infrastructure rehabilitation projects.

Solutions to Practice Problem

Glass fibers are valued for their moderate cost, high tensile strength, and corrosion resistance, making them ideal for reinforcing concrete structures like bridge decks and building facades where durability and affordability are priorities. In contrast, carbon fibers exhibit exceptional mechanical properties, including high tensile strength, stiffness, and fatigue resistance, although they come at a higher cost. They are typically utilized in applications requiring significant load-bearing capacity, such as strengthening columns and beams in high-rise buildings. Aramid fibers offer excellent impact resistance and moderate tensile strength, making them suitable for applications where resistance to dynamic loads is crucial, such as seismic retrofitting of buildings in earthquake-prone regions. Each type of fiber is selected based on its specific properties to enhance the performance and longevity of infrastructure rehabilitation projects, highlighting the versatility and effectiveness of FRP composites in construction and engineering applications.

SUMMARY

- The decline of civil infrastructure, which is vital for societal progress, is a result of aging and insufficient maintenance. Traditional materials such as steel and concrete have certain limitations when it comes to their longevity and durability.
- Fiber-Reinforced polymers (FRP) composites are a highly promising solution for rehabilitation purposes. Offering a range of benefits such as high strength, lightweight corrosion resistance, properties, and comfort of application. In various aspects, FRP composites outperform conventional methods.
- It is crucial to have a comprehensive understanding of the diverse properties and manufacturing procedures associated with the application of composites made of FRP in civil engineering. FRP provides customized features for infrastructure rehabilitation, although there are difficulties related to the absence of uniformity and the need to consider interfaces in application techniques.
- The properties of FRP composites can vary depending on the materials used and the manufacturing processes involved. The importance of durability in civil engineering emphasizes the necessity for a deeper comprehension and anticipation of how materials perform over an extended period.
- The use of FRP composites has brought about significant advancements in rehabilitation projects, encompassing seismic retrofitting and slab repair. These solutions not only offer cost-effective alternatives but also maintain the visual appeal and prolong the lifespan of various structures, including those with historical significance.

MULTIPLE CHOICE QUESTIONS

1. What does FRP stand for in the context of infrastructure rehabilitation?

a. Flexible Repair Products

b. Fiber Reinforced Polymers

c. Ferrocement Reinforcement Panels

d. Formed Resin Plastics

2. Which of the following is NOT a common material used in FRP composites?

a. Carbon fiber

b. Glass fiber

c. Aluminum

d. Aramid fiber

3. What property of FRP composites makes them suitable for infrastructure rehabilitation?

a. High thermal conductivity

b. Low tensile strength

c. Corrosion resistance

d. High density

4. What is the primary purpose of seismic retrofitting in infrastructure rehabilitation?

a. To increase the aesthetic appeal of structures

b. To enhance resistance to earthquakes

c. To improve traffic flow

d. To reduce construction costs

5. Which method is commonly used for seismic retrofitting with FRP composites?

a. Injection grouting

b. Post-tensioning

c. Surface coating

d. Micro-piling

6. What is the primary purpose of strengthening/repair using FRP composites?

a. To decrease load-bearing capacity

b. To improve durability and load-carrying capacity

c. To increase susceptibility to corrosion

d. To reduce the lifespan of structures

REVIEW QUESTIONS

1. Explain the significance of FRP composites in infrastructure rehabilitation, providing examples of their application.
2. Compare and contrast the properties of different types of fibers commonly used in FRP composites for infrastructure rehabilitation.
3. Describe the manufacturing processes involved in producing FRP composites and their respective advantages and limitations.

4. Discuss the role of seismic retrofitting in enhancing the resilience of infrastructure to earthquakes, with a focus on the use of FRP composites.
5. Evaluate the effectiveness of post-tensioning as a method for seismic retrofitting, highlighting its benefits and potential challenges.
6. Analyze the impact of strengthening and repair using FRP composites on the overall performance and service life of infrastructure components, citing real-world case studies where applicable.

Answers to Multiple Questions

1. (b) 2. (c) 3. (c) 4. (b) 5. (b) 6. (b)

REFERENCES

1. Abliz, D., Duan, Y., Steuernagel, L., Xie, L., Li, D., & Ziegmann, G. (2013). Curing methods for advanced polymer composites—A review. *Polymers and Polymer Composites, 21*(6), 341–348. https://doi.org/10.1177/096739111302100604.
2. Ahmad, J., & Ahmad, J. (2009). Introduction to polymer composites. *Machining of Polymer Composites, 8*(2), 1–35. https://doi.org/10.1016/j.composites.2009.02.002.
3. Aisyah, H. A., Paridah, M. T., Sapuan, S. M., Ilyas, R. A., Khalina, A., Nurazzi, N. M., & Lee, C. H. (2021). A comprehensive review on advanced sustainable woven natural fiber polymer composites. *Polymers, 13*(3), 471. https://doi.org/10.3390/polym13030471.
4. Al-Salloum, Y. A., & Almusallam, T. H. (2003). Rehabilitation of the infrastructure using composite materials: Overview and applications. *Journal of King Saud University-Engineering Sciences, 16*(1), 1–20. https://doi.org/10.1016/S1018-3639(18)30120-7.
5. Bakis, C. E., Bank, L. C., Brown, V., Cosenza, E., Davalos, J. F., Lesko, J. J., & Triantafillou, T. C. (2002). Fiber-reinforced polymer composites for construction—State-of-the-art review. *Journal of Composites for Construction, 6*(2), 73–87. https://doi.org/10.1061/(ASCE)1090-0268(2002)6:2(73).
6. Balaguru, P., & Kurtz, S. (1998). Use of inorganic polymer-fiber composites for repair and rehabilitation of infrastructures. In *Proc. Int. Seminar, Workshop, and Exhibition* (Vol. 2, pp. 155–168). https://doi.org/10.1061/9780784403738.017.
7. Bank, L. C. (1993). Properties of FRP reinforcements for concrete. In

Fiber-Reinforced-Plastic (FRP) Reinforcement for Concrete Structures (Vol. 3, pp. 59–86). Elsevier. https://doi.org/10.1016/B978-0-444-89391-5.50010-3.

8. Bedon, C. (2016). Review on the use of FRP composites for facades and building skins. *American Journal of Engineering and Applied Sciences, 9*(3), 713–723. https://doi.org/10.3844/ajeassp.2016.713.723.
9. Berardi, U., & Dembsey, N. (2015). Thermal and fire characteristics of FRP composites for architectural applications. *Polymers, 7*(11), 2276–2289. https://doi.org/10.3390/polym7112276.
10. Cao, S., & Wu, Z. (2008). Tensile properties of FRP composites at elevated and high temperatures. *Journal of Applied Mechanics, 11*, 963–970. https://doi.org/10.1115/1.2943124.
11. Cevik, A., & Cabalar, A. F. (2008). A genetic programming based formulation for the strength enhancement of fiber reinforced polymer confined concrete cylinders. *Journal of Applied Polymer Science, 110*(5), 3087–3095. https://doi.org/10.1002/app.28500.
12. Chen, E. J., & Hsiao, B. S. (1992). The effects of transcrystalline interphase in advanced polymer composites. *Polymer Engineering & Science, 32*(4), 280–286. https://doi.org/10.1002/pen.760320406.
13. Correia, J. R., Almeida, N. M., & Figueira, J. R. (2011). Recycling of FRP composites: Reusing fine GFRP waste in concrete mixtures. *Journal of Cleaner Production, 19*(15), 1745–1753. https://doi.org/10.1016/j.jclepro.2011.06.018.
14. Custódio, J., & Cabral-Fonseca, S. (2013). Advanced fiber-reinforced polymer (FRP) composites for the rehabilitation of timber and concrete structures: Assessing strength and durability. In *Advanced Fiber-Reinforced Polymer (FRP) Composites for Structural Applications* (Vol. 5, pp. 814–882). Woodhead Publishing. https://doi.org/10.1533/9780857096924.4.814.
15. Dai, J. G., Lam, L., & Ueda, T. (2012). Seismic retrofit of square RC columns with polyethylene terephthalate (PET) fibre reinforced polymer composites. *Construction and Building Materials, 27*(1), 206–217. https://doi.org/10.1016/j.conbuildmat.2011.07.030.
16. Das, S. C., & Nizam, M. (2014). Applications of fiber reinforced polymer composites (FRP) in civil engineering. *International Journal of Advanced Structures and Geotechnical Engineering, 3*(3), 299–309. https://doi.org/10.1109/ICACER.2014.6953067.
17. Duic, J., Kenno, S., & Das, S. (2018). Flexural rehabilitation and strengthening of concrete beams with BFRP composite. *Journal of*

Composites for Construction, 22(4), 4018016. https://doi.org/10.1061/(ASCE)CC.1943-5614.0000853.

18. Fahmy, M. F. (2023). Advanced fiber-reinforced polymer composites to enhance seismic response of existing structures. In *Advanced Fiber-Reinforced Polymer (FRP) Composites for Structural Applications* (Vol. 3, pp. 479–539). Woodhead Publishing. https://doi.org/10.1533/9780857096924.4.479.
19. Friedrich, K., & Almajid, A. A. (2013). Manufacturing aspects of advanced polymer composites for automotive applications. *Applied Composite Materials, 20*, 107–128. https://doi.org/10.1007/s10443-012-9268-0.
20. Frigione, M. (2007). Durability aspects of polymer composites used for restoration and rehabilitation of structures. In *Leading-edge Composite Material Research* (Vol. 3, pp. 23–69). Nova Science Publishers. https://doi.org/10.1002/cem.1152.
21. Frketic, J., Dickens, T., & Ramakrishnan, S. (2017). Automated manufacturing and processing of fiber-reinforced polymer (FRP) composites: An additive review of contemporary and modern techniques for advanced materials manufacturing. *Additive Manufacturing, 14*, 69–86. https://doi.org/10.1016/j.addma.2016.12.004.
22. Garden, H. N. (2004). Use of advanced composites in civil engineering infrastructure. *Proceedings of the Institution of Civil Engineers-Structures and Buildings, 157*(6), 357–368. https://doi.org/10.1680/stbu.2004.157.6.357.
23. Ghori, S. W., Siakeng, R., Rasheed, M., Saba, N., & Jawaid, M. (2018). The role of advanced polymer materials in aerospace. In *Sustainable Composites for Aerospace Applications* (Vol. 2, pp. 19–34). Woodhead Publishing. https://doi.org/10.1016/B978-0-08-102831-2.00002-1.
24. Goh, G. D., Yap, Y. L., Agarwala, S., & Yeong, W. Y. (2019). Recent progress in additive manufacturing of fiber reinforced polymer composite. *Advanced Materials Technologies, 4*(1), 1800271. https://doi.org/10.1002/admt.201800271.
25. Gudonis, E., Timinskas, E., Gribniak, V., Kaklauskas, G., Arnautov, A. K., & Tamulėnas, V. (2013). FRP reinforcement for concrete structures: State-of-the-art review of application and design. *Engineering Structures and Technologies, 5*(4), 147–158. https://doi.org/10.3846/20296991.2013.844490.
26. Günaslan, S. E., Karaşin, A., & Öncü, M. E. (2014). Properties of FRP materials for strengthening. *International Journal of Innovative Science, Engineering & Technology, 1*(9), 656–660. https://doi.org/10.5281/zenodo.1078548.

27. Guo, Z., Poot, A. A., & Grijpma, D. W. (2021). Advanced polymer-based composites and structures for biomedical applications. *European Polymer Journal, 149*, 110388. https://doi.org/10.1016/j.eurpolymj.2021.110388.

28. Han, W., Zhou, J., & Shi, Q. (2023). Research progress on enhancement mechanism and mechanical properties of FRP composites reinforced with graphene and carbon nanotubes. *Alexandria Engineering Journal, 64*, 541–579. https://doi.org/10.1016/j.aej.2023.01.015.

29. Hsissou, R., Seghiri, R., Benzekri, Z., Hilali, M., Rafik, M., & Elharfi, A. (2021). Polymer composite materials: A comprehensive review. *Composite Structures, 262*, 113640. https://doi.org/10.1016/j.compstruct.2021.113640.

30. Hulatt, J., Hollaway, L., & Thorne, A. (2003). The use of advanced polymer composites to form an economic structural unit. *Construction and Building Materials, 17*(1), 55–68. https://doi.org/10.1016/S0950-0618(02)00064-3.

31. Iacobucci, R. D., Sheikh, S. A., & Bayrak, O. (2003). Retrofit of square concrete columns with carbon fiber-reinforced polymer for seismic resistance. *Structural Journal, 100*(6), 785–794. https://doi.org/10.14359/14045.

32. Jagadeesh, P., Rangappa, S. M., Suyambulingam, I., Siengchin, S., Puttegowda, M., Binoj, J. S., & Cuadrado, M. M. M. (2023). Drilling characteristics and properties analysis of fiber reinforced polymer composites: A comprehensive review. *Heliyon, 86*(7), 9820–-9825. https://doi.org/10.1016/j.heliyon.2023.e9820.

33. Jose, J. P., & Joseph, K. (2012). Advances in polymer composites: Macro and micro composites–State of the art, new challenges, and opportunities. *Polymer Composites, 8*(3), 1–16. https://doi.org/10.1002/pc.21865.

34. Mosallam, A. (2004). Composites: Construction materials for the new era. In *Advanced Polymer Composites for Structural Applications in Construction* (Vol. 3, pp. 45–58). Woodhead Publishing. https://doi.org/10.1533/9780857096184.1.45.

35. Moy, S. (2013). Fiber-Reinforced Polymer Composite Materials. In *Construction Materials Reference Book* (Vol. 4, pp. 365–376). Routledge. https://doi.org/10.1201/9780203844994.

36. Nanni, A. (2001). Relevant field applications of FRP composites in concrete structures. *Composites in Construction*, Porto, Portugal, October, 10–12. https://doi.org/10.1109/ICACER.2001.939925.

37. Nayana, V., & Kandasubramanian, B. (2020). Advanced polymeric composites via commingling for critical engineering

applications. *Polymer Testing, 91*, 106774. https://doi.org/10.1016/j.polymertesting.2020.106774.

38. Neale, K. W. (2000). FRPs for structural rehabilitation: A survey of recent progress. *Progress in Structural Engineering and Materials, 2*(2), 133–138. https://doi.org/10.1002/pssa.200010022.
39. Pöhler, C., Bachtiar, E. V., Yan, L., & Kasal, B. (2021). Composites for structural strengthening, repair, rehabilitation, and retrofit. In *Composite Materials* (Vol. 3, pp. 205–226). Elsevier. https://doi.org/10.1016/B978-0-12-823610-3.00009-1.
40. Rajak, D. K., Pagar, D. D., Menezes, P. L., & Linul, E. (2019). Fiber-reinforced polymer composites: Manufacturing, properties, and applications. *Polymers, 11*(10), 1667. https://doi.org/10.3390/polym11101667.
41. Rajput, A. S., Sharma, U. K., & Engineer, K. (2019). Seismic retrofitting of corroded RC columns using advanced composite materials. *Engineering Structures, 181*, 35–46. https://doi.org/10.1016/j.engstruct.2018.11.066.
42. Seyhan, E. C., Goksu, C., Uzunhasanoglu, A., & Ilki, A. (2015). Seismic behavior of substandard RC columns retrofitted with embedded aramid fiber reinforced polymer (AFRP) reinforcement. *Polymers, 7*(12), 2535–2557. https://doi.org/10.3390/polym7122535.
43. Silva, P. F., Ereckson, N. J., & Chen, G. D. (2007). Seismic retrofit of bridge joints in central US with carbon fiber-reinforced polymer composites. *ACI Structural Journal, 104*(2), 207. https://doi.org/10.14359/18153.
44. Sonnenschein, R., Gajdosova, K., & Holly, I. (2016). FRP composites and their use in the construction of bridges. *Procedia Engineering, 161*, 477–482. https://doi.org/10.1016/j.proeng.2016.08.711.
45. Toutanji, H. A., & Gomez, W. (1997). Durability characteristics of concrete beams externally bonded with FRP composite sheets. *Cement and Concrete Composites, 19*(4), 351–358. https://doi.org/10.1016/S0958-9465(97)00009-1.
46. Triantafillou, T. C. (2001). Seismic retrofitting of structures with fiber reinforced polymers. *Progress in Structural Engineering and Materials, 3*(1), 57–65. https://doi.org/10.1002/pssa.20001.
47. Truong, B. T., Bui, T. T., Limam, A., Larbi, A. S., Le Nguyen, K., & Michel, M. (2017). Experimental investigations of reinforced concrete beams repaired/reinforced by TRC composites. *Composite Structures, 168*, 826–839. https://doi.org/10.1016/j.compstruct.2017.02.059.
48. Uktamovich, S. B., Yuldashevich, S. A., Rahmonqulovich, A. M., &

Uralbayevich, D. U. (2016). Review of strengthening reinforced concrete beams using CFRP laminate. *European Science Review*, (9–10), 213–215. https://doi.org/10.15587/1729-4061.2016.78623.

49. Uomoto, T., Mutsuyoshi, H., Katsuki, F., & Misra, S. (2002). Use of fiber reinforced polymer composites as reinforcing material for concrete. *Journal of Materials in Civil Engineering, 14*(3), 191–209. https://doi.org/10.1061/(ASCE)0899-1561(2002)14:3(191).
50. Vaghani, M. V., Vasanwala, S. A., & Desai, A. K. (2014). Advanced retrofitting techniques for RC building: A state-of-the-art review. *International Journal of Current Engineering and Technology, 4*(2), 579–584. https://doi.org/10.14741/ijcet/1.3.3.
51. Van Den Einde, L., Zhao, L., & Seible, F. (2003). Use of FRP composites in civil structural applications. *Construction and Building Materials, 17*(6-7), 389–403. https://doi.org/10.1016/S0950-0618(03)00052-2.
52. Wu, C., Xu, F., Wang, H., Liu, H., Yan, F., & Ma, C. (2023). Manufacturing technologies of polymer composites—A review. *Polymers, 15*(3), 712. https://doi.org/10.3390/polym15030712.
53. Xiao, Y. (2004). Applications of FRP composites in concrete columns. *Advances in Structural Engineering, 7*(4), 335–343. https://doi.org/10.1260/1369433042023458.
54. Yu, B., Till, V., & Thomas, K. (2007). Modeling of thermo-physical properties for FRP composites under elevated and high temperature. *Composites Science and Technology, 67*(15–16), 3098–3109. https://doi.org/10.1016/j.compscitech.2007.02.017.
55. Zaman, A., Gutub, S. A., & Wafa, M. A. (2013). A review on FRP composites applications and durability concerns in the construction sector. *Journal of Reinforced Plastics and Composites, 32*(24), 1966–1988. https://doi.org/10.1177/0731684413497884.

CHAPTER 4

RECYCLED AGGREGATES IN GEOTECHNICAL APPLICATIONS

LEARNING OBJECTIVES

At the end of this chapter, readers will be able to:

1. Understand the characteristics of Recycled Aggregates (RAs) for geotechnical applications.
2. Comprehend geotechnical properties of RAs for sustainable construction practices.
3. Know the environmental impact of using RAs in geotechnical engineering.
4. Understand compaction characteristics and shear strength of RAs compared to natural aggregates.
5. Understand the resilient modulus and hydraulic conductivity of recycled aggregates.
6. Understand the particle size distribution and water absorption of recycled aggregates.
7. Explore the unconfined compressive strength and flakiness index of recycled aggregates.

INTRODUCTORY EXAMPLE

GreenTech Construction Company specializes in infrastructure development with a strong focus on environmental responsibility. Recently, the company proposed using recycled aggregates derived from demolished concrete structures to revitalize urban infrastructure. This initiative aims to minimize waste generation while offering a cost-effective and eco-friendly solution for road construction and pavement rehabilitation. The recycled aggregates, processed and supplied by GreenTech, underwent rigorous testing to ensure they meet industry standards. Tests included Particle Size Distribution (PSD) analysis, assessment of material characteristics like compactability and shear strength, and evaluation of hydraulic conductivity and sulphate soundness.

Through vigilant planning and collaboration with engineers and stakeholders, GreenTech successfully integrated recycled aggregates into the construction process. Not only did this initiative reduce the demand for virgin materials, but it also contributed to the reduction of greenhouse gas emissions associated with traditional aggregate production and transportation. By embracing sustainable practices and leveraging the versatility of recycled aggregates, GreenTech Construction Company exemplifies the potential for innovation and environmental responsibility in modern geotechnical engineering projects.

UNIT INTRODUCTION

As infrastructure and urbanization continue to expand, the environment has suffered significant damage due to the construction activities that accompany these developments. This has led to the depletion of natural resources and the generation of large amounts of waste. For example, in 2012, the construction industry was responsible for a significant portion of the total waste generated in the European Union (EU-28). This made it the biggest producer of waste among all economic activities (Figure 4.1). It is worth noting that a significant number of countries within the EU-28 have successfully met the objective of attaining a 70% reuse and recycling rate for safe demolition and construction debris by 2020, as mandated by the Waste Framework Directive (2008/98/EC). Nevertheless, the waste has primarily been utilized for backfilling to facilitate reclamation. Therefore, additional measures must be taken to establish a comprehensive and enduring strategy for sustainability (Cardoso et al., 2016).

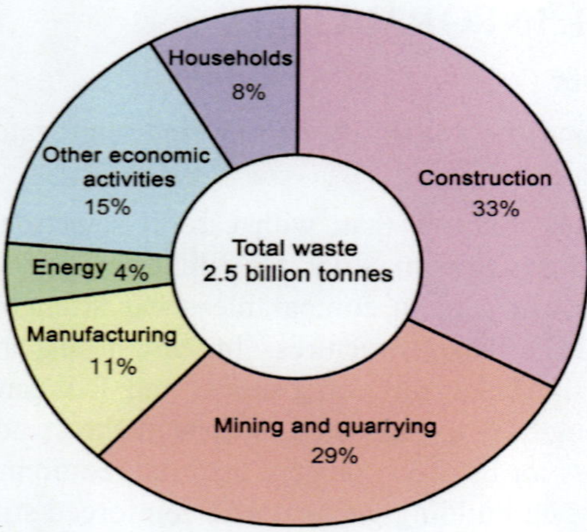

Figure 4.1. *Description of Total breakdown of waste generation in the EU-28.*

Source: Eurostrat 2015, Creative Commons License.

In general, construction and demolition waste are made up of various materials including bricks, excavated soil, paving materials, plastics, glass, wood, concrete, and metals. Several of these materials, especially those containing natural aggregates (NAs), have great potential for being recycled. It is crucial to note the significance of this, considering the substantial annual global production of NA, which is estimated to be around 30 billion tons. Furthermore, it is projected to increase at a rate of 5% per year. In the context of a specific region like Europe, it is worth noting that a significant portion of the total aggregate production is attributed to unbound (40%) and bituminous bound (10%) aggregates (Vieira & Pereira, 2015). Together, these two types of aggregates contribute to approximately half of the overall production. At a global scale, it appears that approximately 15 billion tons of aggregates are utilized for these specific purposes. There is a significant dependence on NAs within geotechnical applications, which calls for a reduction to protect the environment (Arulrajah et al., 2012).

This chapter focuses on evaluating the material features of various types of recycled aggregate, also known as reused aggregates, and attributes of geotechnical engineering. It has been demonstrated that the properties of recycled aggregates are greatly influenced by the type and number of impurities they contain. However, these impurities can be somewhat mitigated through appropriate processing techniques. Recycled aggregate (RA) typically exhibits a higher optimum water concentration and lower maximum dry density when compared to NA because of the presence of cement paste (Sharma et al., 2023).

4.1. GENERAL INFORMATION

Learning Objectives

- Understand the suitability of recycled aggregates in geotechnical applications

Existing research indicates that, with careful selection and compaction, RA can serve as a suitable material for filling in geosynthetic-reinforced embankments. This can lead to comparable shear strength to the materials typically employed in these structures. In certain instances, it has been demonstrated that the links between geogrid and RA can display elevated levels of shear strength. It is suggested to consult the standard specifications of different countries for the necessary fill material requirements, especially in terms of grading, when building geosynthetic reinforced structures. One such specification is BS 8006 (2010). A comprehensive experiment was conducted in Brazil to observe and analyze the performance of geosynthetic-reinforced walls constructed using recycled aggregates (RA) on collapsible foundation soil. The research findings indicated that the deformations of the wall, the durability of the reinforcement, and the strains on the reinforcement all satisfied the required criteria. The results gathered were comparable to walls built with traditional fine materials for backfill and built on a stable foundation (Jayakody et al., 2019).

Furthermore, it is worth noting that the available data also indicates that RA can be a suitable material for various applications such as pipe backfilling material, seawall foundations, gabions, vibro ground improvement works and landfill cover layers. However, it is worth noting that Pererira acknowledged the limited availability of geotechnical information regarding the utilization of Recycled Aggregates (RAs) (Azam & Cameron, 2013).

4.2. THE MATERIAL

Learning Objectives

- Understand the classification and types of recycled aggregates (RA).
- Understand the particle size distribution (PSD) and its implications.
- Know the material characteristics like specific gravity, water absorption, and organic content.

4.2.1 Recycled Aggregate Classification

RA is a broad term that typically refers to aggregates obtained from CDEW (construction, demolition, and excavated waste). Three primary materials were proposed by Silva, which were acquired from CDEW:crushed masonry, crushed concrete, along mixed demolition debris. Additionally, RA has been categorized into four main groups based on its composition. Here is a description of these Duan et al., 2017):

- RCA, also known as Recycled concrete aggregate, is obtained from crushed concrete produced with natural aggregates.
- RMA, also known as Recycled masonry aggregate, is an aggregate that is obtained from masonry rubble, including sand-lime bricks and ceramic bricks.
- MRA, also known as Mixed Recycled Aggregate, is composed primarily of RCA (recycled concrete aggregate) and RMA (Recycled Masonry Aggregate) (Silva et al., 2014).
- CDRA, also known as Construction and Demolition Recycled aggregate is obtained from the waste generated during construction and demolition activities, but unfortunately, it has not undergone adequate processing. Consequently, the composition of the material remains a mystery and may prove challenging to discern (Lux et al., 2023).

Another form of aggregate debris is called RAP also known as Reclaimed Asphalt Pavement, or the aggregate. This aggregate is created through the crushing of bituminous bound combines and is commonly used as a replacement material for geotechnical applications. The primary focus of this chapter is directed towards other areas and does not encompass any pertinent information that is readily available, while also providing a side-by-side analysis of NA and RA (Butler et al., 2014).

4.2.1. Particle Size Distribution (PSD)

PSD of soil significantly impacts its engineering and hydraulic characteristics. Distribution of particle size of recycled aggregates can vary depending on factors such as the demolition procedure used, the intended use of the aggregates, and the type of waste source materials Figure 4.2 displays the PSD of RCA, which has undergone laboratory testing to assess its geotechnical performance. The shaded area in the figure represents over 85% of the complete data (Liu et al., 2020). Additionally, the aggregate classifications provided in BS 5930 (2015) and ASTM D2487 (2006) are included for reference. Figure 4.2 illustrates that most of all RCA samples consisted of a combination of gravel and sand, having a silt content of less than 10% according to ASTM D2487, 2006. For highways being built in the United Kingdom, the use of RCA that meets requirements is required for fill-in structural uses, including Class 6P or 6N as described in Series 600, Earthwork (Highway Agency, 2016a). Figure 4.2 provides a rating indicating that the RCA meets 100% of the criteria, with less than 15% passing through sieve sizes of 80 mm and 63 μm, respectively (Savva et al., 2021). Table 4.1 provides the classification of many types of (RA) recycled aggregates and their corresponding natural aggregates (NA) from 1990 to 2016 in 23 countries. The table includes nearly 250 samples. According to the standards, most of the samples utilized for this evaluation were classified as 'very sandy gravel.' However, according to a different standard, these samples would primarily be classified as 'silty gravel' and 'poorly graded gravel.' A significant majority of the samples in CDRA were processed to create the gravel fraction, with the remaining portion being crushed in the sand fraction. Based on the applicable standards, most RAP samples were categorized as gravel according to BS 5930 (2015), while around fifty percent of the samples had been gravelling and the remaining half were sand according to ASTM D2487 (2006) (Fladvad & Ulvik, 2021).

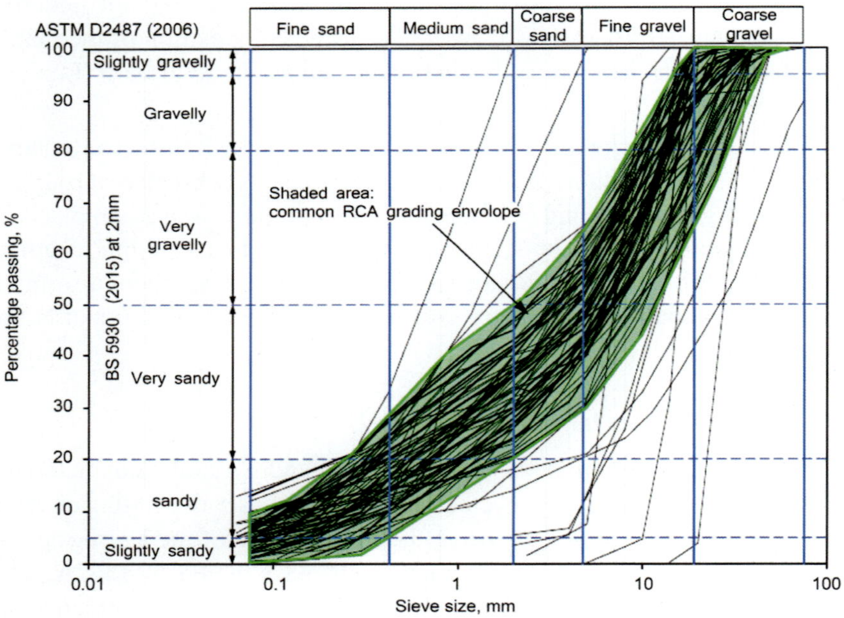

Figure 4.2. *Illustration of particle size distribution of recycled concrete aggregate.*

Source: Bennert L. Creative Commons License.

Table 4.1. *A Comparison of the Classification of Recycled Aggregate Samples using BS and ASTM Standards*

Classification	NA	RCA	RMA	MRA	CDRA	RAP
(a) (BS 5930 2015)						
Sand	2	—	—	—	—	—
Slightly gravelly sand	—	3	—	—	—	—
Gravelly sand	—	—	—	—	—	—
Very gravelly sand	13	3	6	—	25	5
Sand and gravel	10	3	—	—	—	—
Very sandy gravel	60	82	88	94	50	68
Sandy gravel	2	5	—	6	—	27
Slightly sandy gravel	—	—	—	—	—	—
Gravel	13	4	6	—	25	—
(b) ASTM D2487 (2006)						
Well-graded sand	3	1	—	2	—	11
Poorly graded sand	3	16	—	9	—	24
Silty sand	37	2	6	16	29	11
Clayey sand	3	3	—	2	—	—
Well-graded gravel	5	4	6	3	29	20
Poorly graded gravel	5	31	6	25	—	34
Silty gravel	42	43	76	36	29	—
Clayey gravel	—	1	6	7	13	—

Source: Taherkhani. Creative Commons License

4.2.2. Material Characteristics

Table 4.1 provides a list of significant material characteristics that hold great importance in geotechnical studies. Here are the details that are described below.

4.2.2.1. Specific Gravity

The specific gravity of organic materials frequently utilized in geotechnical applications may differ depending on the soil type. On average, it is typically around 2.60-2.75. In geotechnical academic writing, the specific gravity of soil plays a crucial role in determining important material properties like void ratio, porosity, and weight-volume relationship. It is also utilized in various calculations within the field of soil mechanics. Table 4.1 displays an average specific gravity for RCA as 2.42, which is slightly greater than the specific gravity of other types of RA, which ranges from 2.30 to 2.40. The low specific gravity of reused aggregates is mainly attributed to the porous nature of the adhered cement paste present in materials like crushed demolition waste aggregate (CDRA), Recycled Concrete Aggregate (RCA), Recycled Masonry Aggregate (RMA), and Recycled Mixed Aggregate (MRA). Additionally, Recycled Asphalt Pavement (RAP) may have a coating of bitumen with a density that is usually less than 1.10 kg/m3, while CDRA may be contaminated by lightweight materials.

Did You Know?

Integrating recycled aggregates into construction projects can make a significant contribution to achieving LEED (Leadership in Energy and Environmental Design) certification and promoting sustainable building practices.

4.2.2.2. Water Absorption

The porosity of a soil has a strong connection to its water absorption. As indicated in Table 4.1, the rate of water absorption of all the different types of recycled aggregates, except Reclaimed Asphalt Pavement (RAP), is consistently higher than 5%. This is in stark contrast to the water absorbed by Natural Aggregates (NA). It is not surprising that these materials have adhered cement paste, which is porous and therefore absorbs moisture. In geotechnical academic writing, it has been observed that fine RA generally exhibits higher adsorption of water value compared to its coarse RA counterpart. This difference can be attributed to the existence of crushed paste of cement in the form of sand grains. In regards to RAP, residual bitumen coating may hinder complete water absorption. Consequently, the adsorption value of RAP is approximately 2.0%, which is comparatively lower than the absorption values of other types of RA (Rodrigues et al., 2013).

4.2.2.3. Flakiness Index

Particles with a flaky nature can disrupt the arrangement of particles and tend to fracture into smaller fragments when subjected to pressure, such as during the compaction process. The flakiness indices of various types of recycled aggregates typically fall within the range of 14.0%–24.0%. According to Table 4.1, it appears that the materials used have aggregates that have not undergone thorough crushing and sorting. There is a suggestion that RAs that go through primary and secondary crushing procedures often exhibit a low flakiness index.

Typically, NAs exhibit a relatively low flakiness index, except slate which has a higher index due to its natural properties. D. suggests that the flakiness score of aggregates utilized in the sub-base should not exceed 40%. In this scenario, all RAs would fulfill this fundamental criterion (Pourkhorshidi et al., 2020).

4.2.2.4. Los Angeles Abrasion

The L.A. Abrasion Test is commonly employed to assess the various types of aggregates in geotechnical academic writing. The deterioration of standard grading of aggregates is measured by subjecting them to abrasion and impact on a rotating steel drum containing an abrasive charge made of steel balls. The drum is equipped with an internal shelf that raises and lowers the charge and sample about every rotation, creating impact forces. Once the machine has finished running at the specified rpm, the contents are taken out and the percentage of loss is measured. A sample is made by carefully separating it into specific fractions of the necessary masses (Bastidas-Martínez et al., 2019).

The step-by-step procedure of the test is mentioned below:

- The sample of aggregates of a particular size and an abrasive charge are placed within the Los Angeles Abrasion Machines are rotated at a speed of 30-33rpm.
- The sample is carefully extracted and rinsed using a No. 12 (1.70mm) sieve. It is then delicately placed in an oven to undergo the drying process.
- The percentage of mass lost or the difference between the initial mass and the final weight is computed.
- A value of 40 for L.A. Abrasion loss signifies that 40% of the initial sample mass successfully passed through the sieve.

Did You Know?

Recycled aggregates derived from crushed concrete frequently demonstrate superior abrasion resistance compared to natural aggregates, primarily because of the inclusion of cementitious materials.

The L.A. Abrasion, test utilizes larger dried samples along with abrasive charges within a drum equipped with a shelf. The shelf raises and lowers the sample along with the abrasive charge and the span of the drum, leading to a significant impact on the particles of the aggregate. The loss observed in this test is attributed to the fracturing and subsequent dry crushing of the particles (Asadi et al., 2022) (Figure 4.3).

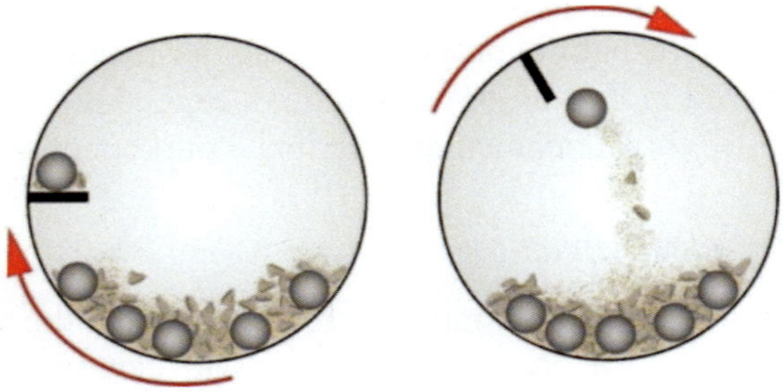

Figure 4.3. *Illustration of the Los Angeles abrasion.*

Source: Gregoire Creative Commons License.

4.2.2.5. Micro-deval Test

The Micro-deval test also evaluates the durability of aggregate, just like the test conducted in Los Angeles abrasion test, but it does so under moist or wet conditions instead of dry ones. It could be beneficial to take into account the impact of moisture in the Micro-Deval test, as this factor has the potential to greatly affect the aggregate's resistance. According to the data in Table 4.1, which is aggregates that have a water absorption of over 5% (such as RCA, RMA, and MRA) tend to have a mean Micro-Deval value that is approximately 20%. This value is higher compared to aggregates with lower water absorption, like NA and RAP, which have a comparable value of around 15% (Ashtiani & Saeed, 2012).

4.2.2.6. Organic Content

Measuring the organic matter of aggregate is crucial as it has a direct impact on its stiffness, three test methods have been employed to determine the natural content of RA. These include loss of mass at 440°C (ASTM D2974, 2014), loss in calcination at 500°C (Vegas et al., 2011), and oxidation through potassium permanganate (UNE 103204, 1993). In geotechnical academic writing, it is important to note that the natural matter content of RA samples is typically higher compared to uncontaminated NA samples (as shown in Table 4.1). This is because the former materials often contain organic impurities like paper and wood (Perkins et al., 2021).

4.2.2.7. Sulphate Content

Sulphate in the soil is considered undesirable due to the possibility of impacting the durability of the concrete foundation. Gypsum products, like drywalls and plasters, in addition to clay bricks, are the primary sources of sulphate in RAs, particularly those derived from construction debris (Głuchowski et al., 2018). According to the data presented in Table 4.1, it is evident that RCA, RMA, and MRA exhibit elevated levels of acid and water-soluble sulphate content, ranging from 0.30% to 1.70%. This indicates that these materials might necessitate additional measures to effectively eliminate the sulphate-bearing impurities. However, unless there is contamination, the amount of sulphate in NA and RAP is typically low, with a mean value of less than 0.2% (Agrela et al., 2014).

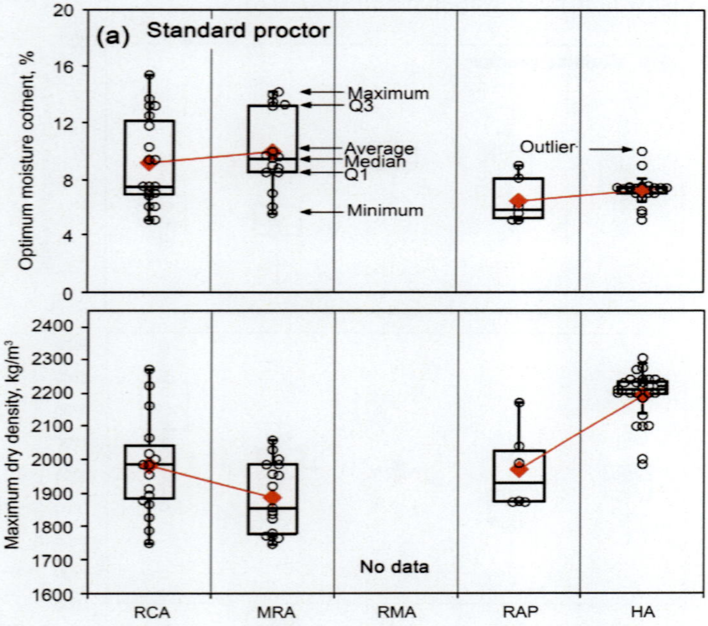

Figure 4.4. *Description of the optimum moisture content and maximum dry density of recycled aggregate and natural aggregate (NA) using the standard proctor tests.*

Source: Cerni and Colagrande, Creative Commons License.

4.3. COMPACTABILITY

Learning Objectives

- Compare compaction characteristics of recycled aggregates with natural aggregates using Proctor tests

In most engineering work, soil is compacted with the application of mechanical energy to rearrange and repack the soil grains to decrease the void ratio between the particles, thereby increasing the shear strength, reducing the settlement under load and making the passage of water harder. Various tests are employed to assess the compactability of soil, with the standard Proctor or modified Proctor tests being widely utilized for this purpose. The main objective is to determine the ideal (OMC) in which the soil achieves its maximum dry density (MDD), suggesting the highest level of compaction. There are distinctions between the two tests, which include variations within the weight of every hammer utilized (2.7 kilograms for one test or 4.9 kilograms for the others) and deviations in the height at which the hammer falls (0.35 cm and also 0.45 cm) (Vishwakarma & Ramachandran, 2018).

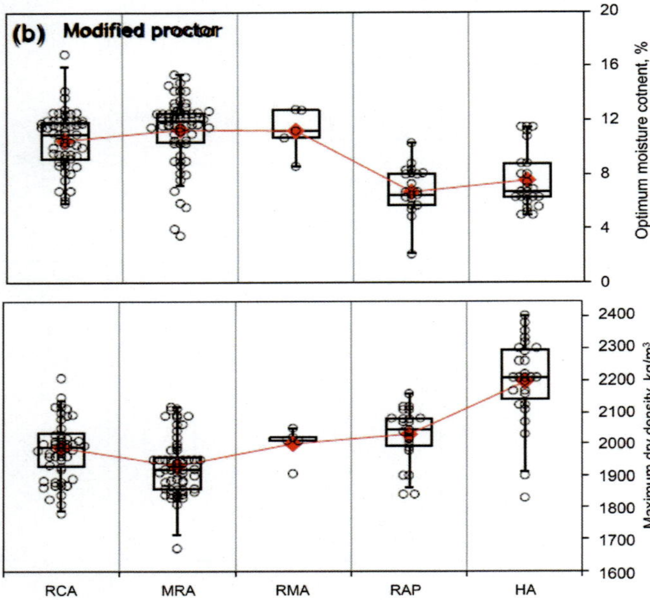

Figure 4.5. *An explanation of the OMC (Optimum Moisture Content) and MDD (Maximum Dry Density) of recycled aggregate and NA (Natural aggregate) through the utilization of modified proctor tests.*

Source: Cerni and Colagrande, Creative Commons License.

The cohesive properties of recycled aggregates (RA) and natural aggregates (NA) were determined by using the conventional Proctor and enhanced Proctor tests. The results are presented in Figures 4.4 and 4.5, respectively. Box-and-whisker plots have been generated for every material to detect any possible outliers. These materials consist of a combination of particles of varying sizes, including gravel and sand. Most of the findings were derived from the improved Proctor test, where RCA and MRA emerged as the most frequently observed data point (Sagario & Ooi, 2011).

Figure 4.4 clearly illustrates why the OMC regarding RAP closely matches that of NA, with a mean OMC value that is approximately 7%, irrespective of the Proctor method used. In contrast, the moisture content of all other recycled aggregates is above 10%, with MRA and RMA having higher moisture content than RCA. The rise in OMC may be explained by the greater absorption of water of RA (Table 4.1) caused by the existence of an adhered paste of cement on the RA or the porous structure of the material, like clay bricks (Islam et al., 2023).

Among the Proctor and standard testing, the average maximum dry density (MDD) of all type values of (RA) recycled aggregates is approximately 2000 kg/, except recycled concrete aggregate (RMA), which has a slightly lower value of 1900 kg/. Nevertheless, the maximum dry density (MDD) values for all recycled aggregate (RA) materials tend to be lower compared to natural aggregate (NA), with an estimated value of approximately 2200 kg/ (Abu Seif, 2015).

4.4. SHEAR STRENGTH

Learning Objectives

- Understand shear strength differences between recycled and natural aggregates in geotechnical engineering

The shear strength of soils is a crucial factor in engineering design, playing a significant role in areas like foundation and slope stability. Two main factors contribute to shear strength. One important factor to consider is the internal angles of friction, which relates to the resistance exhibited by soil particles (Sas et al., 2017). Another important factor to consider is cohesion, which refers to the binding forces that keep the soil particles intact. Figure 4.6 displays the internal angle for cohesion and friction values for RA and NA. These values were obtained through direct and tri-axial shear tests conducted under drained and un-drained conditions. Except for MRA, most of the samples of RA were examined within drained conditions (O'mahony, 1997). The samples had been compacted at the appropriate optimum moisture content and a high percentage of maximum dry density. In geotechnical academic writing, it is worth noting that the internal angles of friction of RA and NA, across both un-drained and drained tests, exhibit striking similarities. These angles typically fall within the range of 40° to 50°, as illustrated in Figure 4.6 (Suddeepong et al., 2020). The value falls within the same range as natural gravel and sand when compacted to a medium to dense level, which typically varies between 36° and 48°. Some of the RA samples, specifically RAP within drained conditions and MRA under un-drained conditions, exhibit a relatively low internal angle of friction. It appears that these materials might have a smoother texture because of the presence of a cement mixture (in this instance of MRA) or a bitumen coating (in the case of RAP). Consequently, there has been a decrease in friction and interlocking capability (Maghool et al., 2019).

Another potential factor contributing to the minimal internal angle for friction in this sample could be poor aggregate grading. Figure 4.6, indicates that certain samples of RA lack cohesion due to their granular composition, as indicated by a pressure of 0 kPa.

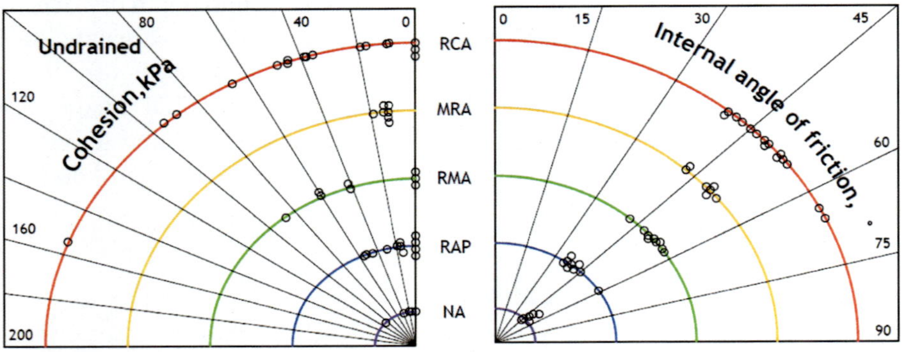

Figure 4.6. *Illustration of Internal angle of friction and cohesion for recycled aggregate materials from direct shear and triaxial tests.*

Source: Molenaar and Niekerk, Creative Commons License.

Past research has demonstrated that the cohesion of recycled aggregates (RA) exhibits significant variation, with values typically not exceeding 100 kPa for materials such as RMA (Recycled Masonry Aggregate) RCA (Recycled Concrete Aggregate), and MRA (Mixed Recycled Aggregate). Similarly, the cohesion of Reclaimed Asphalt Pavement (RAP) generally does not exceed 60 kPa. The bonding between particles is probably facilitated by the existence of an adhered paste of cement or bitumen. Regarding NA, cohesion is linked to the existence of materials that are fine-grained like certain slits and clay (Touahamia et al., 2002).

4.5. UNCONFINED COMPRESSIVE STRENGTH

Learning Objectives

- Understand the role of unconfined compressive strength in assessing recycled aggregates.

The untainted compressive strength analyses evaluate the compressive capacity of cohesive soils when not confined. Additionally, it can be utilized to offer a general assessment of the clay's consistency, whether it is soft or hard (Mohammadinia et al., 2015). The test was also used to examine the self-cementing characteristics of recycled aggregates, particularly recycled concrete aggregates, because of the cement paste that is attached to them. Figure 4.7 displays the compressive strength in an unconfined state of untreated NA and RA (without using cement materials) over 1-28 days (Saberian & Li, 2021).

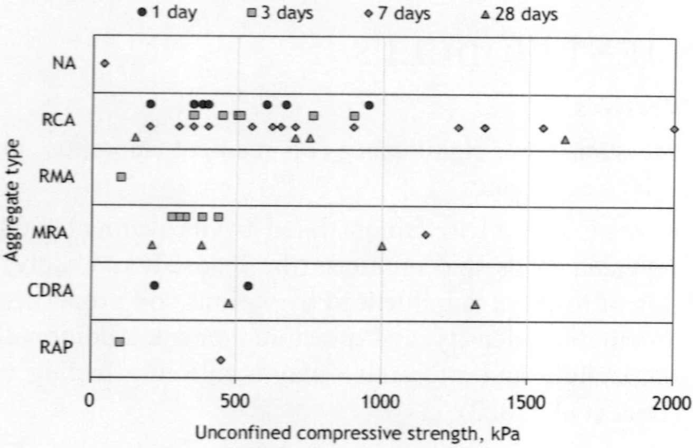

Figure 4.7. Illustration of Unconfined compressive strength of untreated recycled aggregate and natural aggregate.

Source: Cameron, Creative Commons License.

The strength of the RA specimens is primarily influenced by the quality and quantity of the adhered cement paste. This binding effect is achieved through the rehydration process, as the specimens did not undergo treatment with any cementitious materials. Figure 4.7, shows that RCA generally exhibits higher unconfined compressive strengths compared to other recycled aggregate materials at a given age. It is not surprising that RCA, which comes from crushed concrete, may contain a higher amount of cement paste residue (Yaghoubi et al., 2017).

Additionally, certain studies have noted that the strength of RCA samples showed an increase over time. In a particular instance, it is also demonstrated that specimens made of Recycled Concrete aggregates (RCA) demonstrate continuous strength improvement over 180 days. Figure 4.7, concludes that regardless of their age, it is highly improbable for the unrestricted compressive force of RCA to surpass 1000 kPa (Gu et al., 2023). However, it is anticipated that the unconfined compressive strength of alternative materials made from recycled aggregates (RA) will be less than 500 kPa. It has been noted that the self-cementing features of recycled aggregates can alter their intended engineering properties, transforming an unbound material into a bound material. Thus, it has been proposed that utilizing non-reactive aggregates from recycled materials (specifically, RMA produced from waste bricks) may be a more appropriate option in this scenario (Karkush & Yassin, 2019).

4.6. RESILIENT MODULUS

Learning Objectives

- Understand the significance of resilient modulus in assessing recycled aggregates.

The stiffness of the soil is demonstrated by its resilient modulus when subjected to repeated loads that mimic traffic loads. It is widely recognized that the modulus of resilient is influenced by various soil properties, including particle size distribution, density, and moisture content. Additionally, external factors like temperature and stress also play a role in affecting the resilient modulus (Oskooei et al., 2022).

Figure 4.8 displays the resilient moduli of Recycled Concrete Aggregates (RCA) and Natural Aggregates (NA) plotted contrary to their corresponding bulk stress. The bulk stress is determined using an equation provided below, in cases where this information is not readily available. In addition, the categorization of bulk stress data into 100 kPa increments to generate a box-and-whiskers plot for every group (Bozyurt et al., 2012):

When considering the stress values, it is important to note the deviator stress () and the confinement stress (). These values are measured in kilopascals (kPa).

According to Figure 4.8, the resilient moduli for both Recycled Concrete Aggregate (RCA) and Natural Aggregate (NA) are observed to increase with an increase in bulk stress. However, it is worth noting that RCA generally exhibits a greater resilient modulus compared to NA. Some researchers have attributed this to the greater angularity and rougher surface texture of RCA, although it is not commonly explained. On the other hand, it was discovered that the Resilient Modulus of RCA is lower compared to NA. This indicates that the angularity of NA is more effective in distributing the load than RCA. In terms of material properties, the cement paste that adheres to the Recycled Concrete Aggregate (RCA) is porous along with a lower modulus of elasticity. The resistance of the RCA against deformation is weakened, leading to a comparatively small resilient modulus. Therefore, additional research is necessary to elucidate the impact of the properties of both Natural Aggregates (NA) and Recycled Concrete Aggregates (RCA) on their resilient modulus. When conducting additional tests, it is crucial to maintain the same Particle Size Distribution (PSD) for both materials. This will prevent any potential discrepancies that may arise from differences in particle packing between the two materials (Xiao et al., 2022).

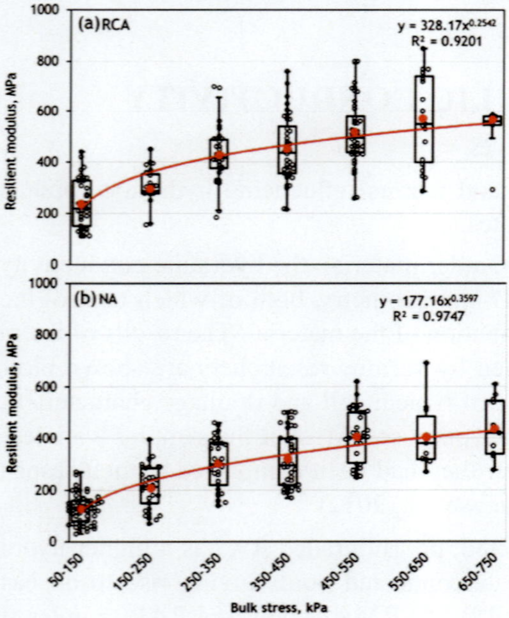

Figure 4.8. *Illustration of Resilient modulus results for (a) recycled concrete aggregate (RCA) and (b) natural aggregate (NA).*

Source: Boyzurt, Creative Commons License.

However, certain characteristics have been discovered to impact the resilient modulus of RCA (Recycled Concrete Aggregates):

- Water Content: The resilient modulus of RCA decreases when water content increases.
- As the curing age progresses, the resilient modulus for RCA experiences a gradual increase.

- Freeze-Thaw cycle: According to research findings, it has been observed that the resilient modulus of RCA tends to increase as the freeze-thaw cycles progress. This increase is believed to be a result of the rehydration of the cement particles present in RCA.
- Temperature: The resilient modulus of Recycled Concrete Aggregates (RCA) shows minimal variation when tested at temperatures ranging from 7°C to 50°C.
- Compaction method: Resilient moduli of Recycled Concrete Aggregates (RCA) generated by vibro compression and vibrating hammer methods exhibit a high degree of similarity (Diagne et al., 2015).

4.7. HYDRAULIC CONDUCTIVITY

Learning Objectives

- Understand factors influencing hydraulic conductivity in recycled aggregates.

As with any granular material, the hydraulic conductivity of a soil may be affected by its PSD and geometry, both of which control the particle packing and pore size distribution of the material. The results of hydraulic conductivity tests of RA obtained by various researchers are shown plotted in Figure 4.9, together with those of typical soil and drainage characteristics as provided by Barnes, for reference purposes. Most of the samples were tested using a falling head apparatus after they had been compacted to their respective OMC and at 98% MDD (Nokkaew et al., 2012).

As to be expected, poorly graded RA has a higher hydraulic conductivity value due to its interconnected voids giving rise to an easy flow of water. Although the tested RCA, RMA, MRA and RAP samples were a mixture of gravel and sand, their hydraulic conductivity values were similar to those of silts (poor drainage) and stratified clay/silt deposits (practically impervious) (Yu et al., 2023). This is perhaps due to the adhered cement paste or bitumen residue of RA breaking into smaller pieces during compaction, and such fine particles filling the voids and thereby making the material denser and less permeable (Abdulnafaa et al., 2021).

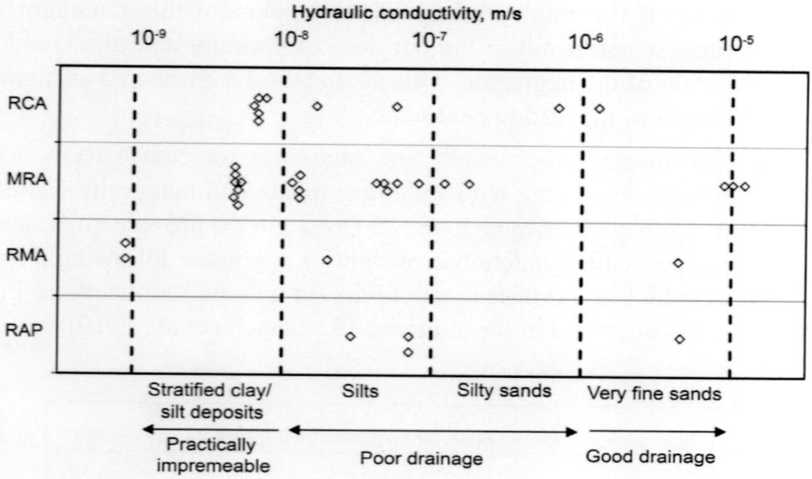

Figure 4.9. *Illustration of Hydraulic conductivity data for different recycled aggregate materials.*

Source: Diagne, Creative Commons License.

MRA known as (Mixed Recycled Aggregate).

RAP known as (Reclaimed Asphalt Pavement).

RCA known as (Recycled Concrete Aggregate).

RMA known as (Recycled Masonry Aggregate) (Rahman et al., 2015).

4.8. SULPHATE SOUNDNESS

Learning Objectives

- Understand how sulphate soundness values of recycled aggregates compare to natural aggregates do.

The durability of aggregate in geotechnical academic writing is a crucial property that affects its resistance to degradation caused by environmental factors. This property plays a significant role in modifying the stiffness and shear strength of aggregates, ultimately leading to potential failure. Tests for sulphate soundness, like those outlined in BS EN 1367-2 (2009) and ASTM C88 (2013), evaluate the durability of aggregates by measuring weight loss. This process involves exposing the aggregates to multiple rounds of immersing in a solution that is completely saturated with magnesium sulphate or sodium sulphate, and oven dry them. When it comes to geotechnical academic writing, one important factor to consider is the buildup of crystals of sulphate within the aggregate's pores. This process can result in the build-up of pressure and ultimately lead to the degradation of the aggregate. This degradation is measured as a gradual decrease in weight as the testing continues.

Figure 4.10 displays the weight loss outcomes for coarse RCA, MRA, and NA obtained from testing with sodium sulphate and magnesium sulphate. The Highway Agency (2016b) and ASTM D692 (1995) provide guidelines on the maximum acceptable reduction in weight of aggregate following a sulfate soundness test, which determines its suitability for unbound applications. Figure 4.10 displays this information for reference (Rattanachu et al., 2019).

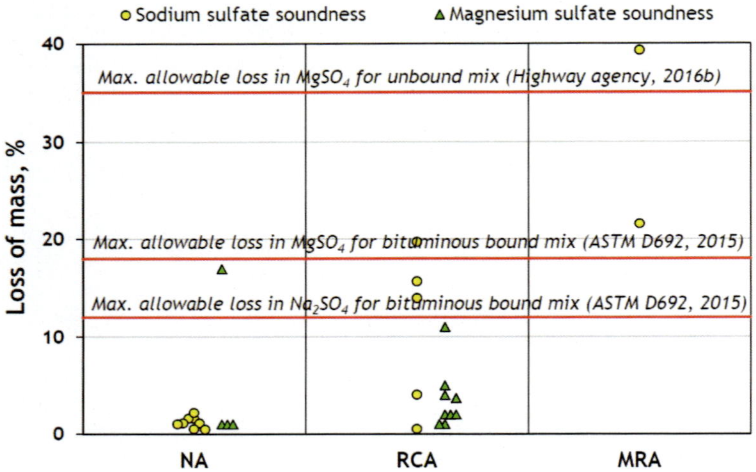

Figure 4.10. *Illustration of Sodium sulphate and magnesium sulphate soundness of coarse recycled aggregate and natural aggregate (NA). MRA, mixed recycled aggregate; RCA, recycled concrete aggregate.*

Source: Bestgen, Creative Commons License.

The sulphate soundness value of NA tends to be quite low, usually not exceeding 3% (as shown in Figure 4.10). However, there was one exceptional case where the magnesium sulphate soundness of limestone reached as high as 17%. On the other hand, the sulphate soundness rating for recycled concrete aggregates, also known as RCA, is generally higher than that of natural aggregates (NA). This is because RCA contains porous and weak cement paste that is attached to it. It is worth noting that a majority of the RA samples demonstrate a rating of sulphate soundness is less than 10% (Figure 4.10) (Bilal & Khan, 2019). This suggests that the material exhibits impressive durability against weathering. According to Figure 4.10, it can be observed that MRA exhibits the greatest loss of mass values during the sulphate soundness test. This is due to the presence of 25% to 50% crushed bricks of clay in the material, which ultimately leads to complete disintegration after the tests. In general, the findings in Figure 4.10 indicate that RCA may be a better choice for bituminous bound and unbound mixes in road pavements, as compared to MRA. This is especially true when the amount of weak elements, like adhered cement mixture, is comparatively low (Bassani et al., 2017).

4.9. ENVIRONMENTAL IMPACT

Learning Objectives

- Assess environmental impact and potential contamination risks in recycled aggregates.

When RA is used instead of NA in geotechnical applications, it is unlikely to pose a major environmental hazard, as long as it is not contaminated. This is because the material comes from natural resources. On the other hand, it is worth noting that RCA, MRA, and RMA exhibited a slightly acidic to acidic composition (with a pH of 7.5 to 11). One possible explanation for this is the existence of leftover cement paste material, which stands in contrast to the almost neutral values seen in the case of NA and RAP.

Minimizing the chance of potential contaminants in RA involves implementing a thorough sorting process to remove any foreign materials. However, it is important to note that CDRA, typically a waste matter coming from building and demolition procedures, must undergo a thorough and regular quality check when designated for use in geotechnical applications.

Yet researchers have performed chemical leachate analyses on different types of aggregates, such as MRA, RMA, RCA, and RAP, to assess the release of potentially harmful elements like Ba, Cu, Cr, Mo Ni, Se, Sb, and Zn. The analyses were performed using BS EN 12457-3, a standard widely recognized in the field. All elements released from the RA were found to have concentrations below the waste acceptable limit for inert waste in the United Kingdom. However, certain RMA and MRA samples showed a high level of Cr. Once again, it is important to emphasize the importance of precise categorization when working with the material (Inui et al., 2009).

Practice Problem

KARVAAN Construction Company specializes in sustainable infrastructure development and construction solutions. KARVAAN Construction is exploring the use of recycled aggregates (RA) in their geotechnical projects as a means to promote sustainability and minimize the impact on the environment. Nevertheless, there is some uncertainty surrounding the environmental consequences of utilizing RA in contrast to NA, specifically concerning pH levels and the leaching of contaminants.

Solutions to Practice Problem

To tackle this issue, KARVAAN Construction can carry out a thorough environmental impact assessment. First, researchers can gather samples of both recycled aggregates (RA) and natural aggregates (NA) from their chosen project locations. And can then perform pH testing to identify any deviations from a neutral pH. Through the analysis of pH levels and comparison with established environmental standards, researchers can evaluate the potential dangers of alkalinity linked to RA. In addition, KARVAAN Company can conduct chemical leachate analyses of RA samples, specifically Recycled Masonry Aggregate (RMA), to identify any potential presence of contaminants, such as chromium. These findings can be evaluated against regulatory guidelines to assess compliance and determine any possible risks related to environmental contamination. These findings can help KARVAAN Construction implement effective mitigation measures, including sorting and quality control procedures, to ensure the responsible use of RA in their geotechnical projects. This approach promotes sustainability and reduces environmental impact.

SUMMARY

- The chapter offers a thorough examination of the utilization of recycled aggregates (RA) in geotechnical uses, with a specific focus on comparing them to natural aggregates (NA).
- It emphasizes the potential of RA as a suitable material for use in geosynthetic-reinforced embankments, demonstrating shear strength that is comparable to traditional materials and even greater shear strength at geogrid interfaces.
- The classification of RA types, including recycled concrete aggregate (RCA), recycled masonry aggregate (RMA), and mixed recycled aggregate (MRA), is detailed, along with their particle size distribution and material characteristics such as specific gravity, water absorption, and flakiness index.
- The paper discusses the results of various tests, including unconfined compressive strength, resilient modulus, hydraulic conductivity, compactibility, and shear strength. These tests shed light on the behavior of RA under different conditions.
- Environmental considerations include RA's alkaline composition and the importance of proper sorting to minimize contamination. Overall, while RA shows promise as a sustainable alternative to NA, further research and quality control measures are needed to fully understand and optimize its performance in geotechnical applications.

MULTIPLE CHOICE QUESTIONS

1. Which of the following is NOT a type of recycled aggregate?

a. Crushed concrete

b. Reclaimed asphalt pavement (RAP)

c. Recycled plastic

d. Recycled glass

2. What aspect of recycled aggregates influences their ability to achieve desired density in engineered fills?

a. Particle size distribution

b. Material density

c. Absorption capacity

d. Abrasion resistance

3. Which testing method is used to assess the load-bearing capacity of recycled aggregates?

a. Unconfined compressive strength

b. Resilient modulus

c. Hydraulic conductivity

d. Compaction test

4. Resilient modulus testing is primarily employed for evaluating:

a. Stiffness and deformation characteristics

b. Hydraulic conductivity

c. Environmental impact

d. Material density

5. What geotechnical property does hydraulic conductivity assess?

a. Drainage and permeability

b. Load-bearing capacity

c. Compaction characteristics

d. Resilient modulus

6. Using recycled aggregates in geotechnical applications helps in:

a. Increasing waste disposal

b. Depleting natural resources

c. Minimizing carbon footprint

d. Enhancing environmental pollution

REVIEW QUESTIONS

1. What are the main types and classifications of recycled aggregates?
2. How does the particle size distribution of recycled aggregates compare to that of natural aggregates?
3. What are the key material characteristics of recycled aggregates in geotechnical applications?
4. How is the compactability of soil affected by the use of recycled aggregates?
5. What factors influence the shear strength of soils containing recycled aggregates?
6. How does the unconfined compressive strength of recycled aggregates compare to that of natural aggregates?

7. What role does the resilient modulus play in evaluating the performance of recycled aggregates in geotechnical applications?
8. How does the hydraulic conductivity of soil containing recycled aggregates differ from soil with natural aggregates, and what factors influence this difference?

Answers to Multiple Questions

1. (c) 2. (a) 3. (a) 4. (a) 5. (a) 6. (c)

REFERENCES

1. Abdulnafaa, M., Awad, M., Cabalar, A., Akbulut, N., & Ozufacik, B. (2021). Hydraulic conductivity and undrained shear strength of clay-construction and demolition solid waste materials mixtures. *Soils and Rocks, 44*, 2021062221. https://doi.org/10.28927/SR.4420211062221.
2. Abu Seif, E. S. S. (2015). Geotechnical approach to evaluate natural fine aggregates concrete strength, Sohag, Governorate, Upper Egypt. *Arabian Journal of Geosciences, 8*(9), 7565–7575. https://doi.org/10.1007/s12517-015-1953-4.
3. Arulrajah, A., Piratheepan, J., Ali, M. M. Y., & Bo, M. W. (2012). Geotechnical properties of recycled concrete aggregate in pavement sub-base applications. *Geotechnical Testing Journal, 35*(5), 743–751. https://doi.org/10.1520/GTJ20120039.
4. Asadi, M., TaghaviGhalesari, A., & Kumar, S. (2022). Machine learning techniques for estimation of Los Angeles abrasion value of rock aggregates. *European Journal of Environmental and Civil Engineering, 26*(3), 964–977. https://doi.org/10.1080/19648189.2020.1760841.
5. Ashtiani, R., & Saeed, A. (2012). Laboratory performance characterization of pavements incorporating recycled materials. In *GeoCongress 2012: State of the Art and Practice in Geotechnical Engineering* (Vol. 3, pp. 1582–1591). https://doi.org/10.1061/9780784412347.165.
6. Azam, A. M., & Cameron, D. A. (2013). Geotechnical properties of blends of recycled clay masonry and recycled concrete aggregates in unbound pavement construction. *Journal of Materials in Civil Engineering, 25*(6), 788–798. https://doi.org/10.1061/(ASCE)MT.1943-5533.0000618.
7. Cardoso, R., Silva, R. V., de Brito, J., & Dhir, R. (2016). Use of recycled aggregates from construction and demolition waste in geotechnical applications: A literature review. *Waste Management, 49*, 131–145. https://doi.org/10.1016/j.wasman.2015.10.014.

8. Diagne, M., Tinjum, J. M., & Nokkaew, K. (2015). The effects of recycled clay brick content on the engineering properties, weathering durability, and resilient modulus of recycled concrete aggregate. *Transportation Geotechnics, 3*, 15–23. https://doi.org/10.1016/j.trgeo.2015.01.002.
9. Duan, Z., Poon, C. S., & Xiao, J. (2017). Using artificial neural networks to assess the applicability of recycled aggregate classification by different specifications. *Materials and Structures, 50*, 1–14. https://doi.org/10.1617/s11527-017-1102-6.
10. Fladvad, M., & Ulvik, A. (2021). Large-size aggregates for road construction—A review of standard specifications and test methods. *Bulletin of Engineering Geology and the Environment, 80*(12), 8847–8859. https://doi.org/10.1007/s10064-021-02260-5.
11. Głuchowski, A., Sas, W., Dzięcioł, J., Soból, E., & Szymański, A. (2018). Permeability and leaching properties of recycled concrete aggregate as an emerging material in civil engineering. *Applied Sciences, 9*(1), 81. https://doi.org/10.3390/app9010081.
12. Gu, M., Cai, X., Mo, H., & Wang, Q. (2023). Unconfined compressive behavior of column with geogrid-encased recycled aggregate: Test and simulation. *Construction and Building Materials, 400*, 132528. https://doi.org/10.1016/j.conbuildmat.2023.132528.
13. Islam, S., Islam, J., Hoque, N. M. R., & Hasan, K. (2023). Improving geotechnical properties of soil of hillock slope using crushed recycled concrete aggregates. *Journal of Engineering Research, 7*(3), 100139. https://doi.org/10.36909/jer.100139.
14. Jayakody, S., Gallage, C., & Ramanujam, J. (2019). Effects of reclaimed asphalt materials on geotechnical characteristics of recycled concrete aggregates as a pavement material. *Road Materials and Pavement Design, 20*(4), 754–772. https://doi.org/10.1080/14680629.2018.1524715.
15. Karkush, M. O., & Yassin, S. (2019). Improvement of geotechnical properties of cohesive soil using crushed concrete. *Civil Engineering Journal, 5*(10), 2110–2119. https://doi.org/10.28991/cej-2019-03091344.
16. Lux, J., Hoong, J. D. L. H., Mahieux, P. Y., & Turcry, P. (2023). Classification and estimation of the mass composition of recycled aggregates by deep neural networks. *Computers in Industry, 148*, 103889. https://doi.org/10.1016/j.compind.2022.103889.
17. Maghool, F., Arulrajah, A., Suksiripattanapong, C., Horpibulsuk, S., & Mohajerani, A. (2019). Geotechnical properties of steel slag aggregates: Shear strength and stiffness. *Soils and Foundations, 59*(5), 1591–1601. https://doi.org/10.1016/j.sandf.2019.07.008.

18. Naeini, M., Mohammadinia, A., Arulrajah, A., & Horpibulsuk, S. (2021). Recycled glass blends with recycled concrete aggregates in sustainable railway geotechnics. *Sustainability, 13*(5), 2463. https://doi.org/10.3390/su13052463.
19. Nokkaew, K., Tinjum, J. M., & Benson, C. H. (2012). Hydraulic properties of recycled asphalt pavement and recycled concrete aggregate. In *GeoCongress 2012: State of the Art and Practice in Geotechnical Engineering* (Vol. 3, pp. 1476–1485). https://doi.org/10.1061/9780784412347.166.
20. O'mahony, M. M. (1997). An analysis of the shear strength of recycled aggregates. *Materials and Structures, 30*, 599–606. https://doi.org/10.1007/BF02473377.
21. Perkins, L., Royal, A. C., Jefferson, I., & Hills, C. D. (2021). The use of recycled and secondary aggregates to achieve a circular economy within geotechnical engineering. *Geotechnics, 1*(2), 416–438. https://doi.org/10.1016/j.geotech.2021.100003.
22. Pourkhorshidi, S., Sangiorgi, C., Torreggiani, D., & Tassinari, P. (2020). Using recycled aggregates from construction and demolition waste in unbound layers of pavements. *Sustainability, 12*(22), 9386. https://doi.org/10.3390/su12229386.
23. Rodrigues, F., Evangelista, L., & Brito, J. D. (2013). A new method to determine the density and water absorption of fine recycled aggregates. *Materials Research, 16*, 1045–1051. https://doi.org/10.1590/S1980-53732013000400013.
24. Saberian, M., & Li, J. (2021). Effect of freeze–Thaw cycles on the resilient moduli and unconfined compressive strength of rubberized recycled concrete aggregate as pavement base/subbase. *Transportation Geotechnics, 27*, 100477. https://doi.org/10.1016/j.trgeo.2021.100477.
25. Sagario, M. L., & Ooi, P. S. (2011). Assessing the compatibility of recycled concrete aggregate. *Transportation Research Record, 2253*(1), 10–21. https://doi.org/10.3141/2253-02.
26. Sas, W., Głuchowski, A., & Szymański, A. (2017). The geotechnical properties of recycled concrete aggregate with addition of rubber chips during cyclic loading. *GEOMATE Journal, 12*(29), 25–32. https://doi.org/10.21660/2017.29.34138.
27. Savva, P., Ioannou, S., Oikonomopoulou, K., Nicolaides, D., & Petrou, M. F. (2021). A mechanical treatment method for recycled aggregates and its effect on recycled aggregate-based concrete. *Materials, 14*(9), 2186. https://doi.org/10.3390/ma14092186.

28. Sharma, A., Shrivastava, N., & Lohar, J. (2023). Assessment of geotechnical and geo-environmental behaviour of recycled concrete aggregates, recycled brick aggregates and their blends. *Cleaner Materials, 7*, 100171. https://doi.org/10.1016/j.clema.2023.100171.
29. Touahamia, M., Sivakumar, V., & McKelvey, D. (2002). Shear strength of reinforced-recycled material. *Construction and Building Materials, 16*(6), 331–339. https://doi.org/10.1016/S0950-0618(02)00009-7.
30. Vieira, C. S., & Pereira, P. M. (2015). Use of recycled construction and demolition materials in geotechnical applications: A review. *Resources, Conservation and Recycling, 103*, 192–204. https://doi.org/10.1016/j.resconrec.2015.07.009.
31. Vishwakarma, V., & Ramachandran, D. (2018). Green concrete mix using solid waste and nanoparticles as alternatives–A review. *Construction and Building Materials, 162*, 96–103. https://doi.org/10.1016/j.conbuildmat.2017.12.047.
32. Yu, M., Gui, Y., & Laguna, R. (2023). Hydraulic conductivity characteristics of a clayey soil incorporating recycled rubber and glass granules. *Water, 15*(11), 2028. https://doi.org/10.3390/w15112028.

CHAPTER 5

ADVANCED CONCRETE FOR CIVIL INFRASTRUCTURE

LEARNING OBJECTIVES

At the end of this chapter, readers will be able to:

1. Understand the importance of quality control in concrete production.
2. Understand the chemistry and production process of Portland cement.
3. Identify the properties and uses of supplementary cementing materials.
4. Understand principles and properties of high-performance concretes.
5. Know factors affecting durability and strength in concrete materials.
6. Comprehend advancements in concrete technology for diverse applications.
7. Recognize strategies for sustainable and environmentally friendly concrete production.

INTRODUCTORY EXAMPLE

Concrete plays a vital role in the construction industry, with advancements continually shaping its properties and applications. Consider a project by "Advanced Construction Solutions Inc.," a leading construction company renowned for its innovative approach to infrastructure development.

In a recent project, Advanced Construction Solutions was tasked with building a high-rise residential complex in an urban setting. To meet the project's demanding requirements for strength, durability, and sustainability, the company opted for modern advanced concrete solutions. By utilizing ultra-high-strength concrete for the building's structural elements, Advanced Construction Solutions ensured that the complex could withstand both environmental stresses and the test of time. Additionally, the incorporation of fiber-reinforced concrete enhanced the structure's resilience against seismic activity, providing added safety for its future residents.

Moreover, the company embraced self-compacting concrete for its ease of placement and superior finish, streamlining the construction process while maintaining high standards of quality. Through the strategic use of polymer-modified concrete, Advanced Construction Solutions further minimized environmental impact and maximized resource efficiency, aligning with their commitment to sustainable construction practices. This project exemplifies the importance of modern advanced concrete in meeting the evolving needs of civil infrastructure. As construction companies like Advanced Construction Solutions continue to push the boundaries of concrete technology, the possibilities for safer, more resilient, and environmentally friendly structures are endless.

UNIT INTRODUCTION

Cementitious materials have a long history as construction materials and continue to be widely used today. Their production surpasses that of steel, timber, asphalt, and other construction materials. Their history can be traced back thousands of years, as evidenced by the discovery of a mortar floor in Israel. Comparable lime mortars, created through the process of calcining limestone, were widely utilized in various regions of the Middle East. Impure gypsum was calcined to prepare gypsum-based mortars, which were utilized in ancient Egypt for various construction projects, including the Pyramid of Cheops (~3000 B.C.). However, both these kinds of cementing materials lack hydraulic properties, meaning not get harder when exposed to water (Garden, 2004).

In later times, the Greeks along with the Romans successfully developed hydraulic lime mortars by incorporating extremely fine volcanic ash.

Specifically, the Romans utilized a substance known as pozzolana, which was discovered near the small town of Pozzuoli, near Mt Vesuvius. Vitruvius wrote about this material around 27 B.C.

Additionally, there also exists a type of powder that yields remarkable outcomes due to its natural properties. It can be located in the vicinity of area Baiea, and within the region that belongs to the surrounding towns near Mt Vesuvius. This material, when combined with lime and debris, not only enhances the structural integrity of various types of buildings but also solidifies even when used to construct piers in marine environments (Hollaway, 2003).

These types of materials were widely utilized in hydraulic structures, including piers and sea walls. They were also utilized in the creation of a type of concrete. The Pantheon, a structure dating about the second part of the 2nd century A.D., is widely regarded as one of the most well-preserved Roman structures. It was primarily constructed using concrete, showcasing the remarkable durability of this material. The forty-four m diameter dome had been constructed by pouring concrete in ribbed sections and allowing it to solidify. Furthermore, to decrease the overall weight, a lightweight aggregate made from crushing a porous volcanic material (pumice) was utilized (

During the Middle Ages, the standard of cementing materials experienced a decline. However, following the 14th century A.D., high-quality mortars were rediscovered. Subsequently, in the second half of the 18th century, studies were initiated, ultimately leading to the development of what is known as 'modern' Portland cement. In 1756, John Smeaton was tasked with the reconstruction of the Eddystone Lighthouse in Cornwall, England. During this project, he conducted a series of experiments using various types of limes and pozzolans. Through his research, he found that limestone with a significant amount of clayey material was the most suitable for this particular application. Although it was not exactly Portland cement, it marked a significant advancement compared to the slaked limes that were being used at that time (Hollaway, 2002).

However, for those unfamiliar with the subject, concrete may appear to be a straightforward substance: There are several important aspects to consider when working with the materials used in civil infrastructure construction. By combining cement, water, and aggregates, a chemical reaction occurs that results in the hardening of the mixture. However, there is much more to learn about all of the details involved in utilizing concrete for civil infrastructure projects. Truly, the apparent simplicity of concrete has contributed to its extensive use as a material. The actuality is quite distinct. Contemporary concretes, contrary to common misconceptions, are indeed advanced materials. Our capacity to develop cutting-edge concretes, with precise control over their flow characteristics, incredibly strong compressive strength, impressive tensile strength, remarkable toughness, and exceptional long-term durability, relies on our deep comprehension of the underlying chemical and mechanical characteristics of the concrete. Additionally, our expertise in manipulating the material's microstructure, and in some cases even its nanostructure, plays a crucial role. To achieve such remarkable feats of engineering, it is crucial to have a deep understanding of the properties and applications of concrete. In the upcoming sections, the exploration of some of the latest advancements in the manufacturing and application of advanced concrete (Bautista-Gutierrez et al., 2019).

5.1. MODERN ADVANCED CONCRETE

Learning Objectives

- Understand advanced concrete's concept over conventional mixtures.

High-performance concrete, also known as advanced concrete, is a type of concrete that goes beyond the usual requirements for durability and uniformity. It is designed to meet specific criteria that are rarely feasible with regular materials and standard procedures for mixing, placing, and curing. In simpler terms, advanced concrete refers to a type of concrete that surpasses the typical concrete produced in various aspects (Kioumarsi & Plevris, 2024). The definition mentioned above is subject to change over time, much like the nature of concrete. What may have been considered as cutting-edge concrete a century ago would appear quite ordinary by today's standards. Similarly, the so-called 'advanced' concretes of today will likely be regarded as primitive in the future (Feizbahr & Pourzanjani, 2024).

5.2. MATERIALS

Learning Objectives

- Understand the essential components and considerations in producing high-performance concrete.
- Recognize the importance of quality control throughout the concrete-making process.
- Comprehend the roles and properties of Portland cement, aggregates, chemical admixtures, and mineral admixtures.
- Know the influence of supplementary cementing materials like fly ash and silica fume.

Before delving into the characteristics of modern concrete options, it is important to provide an overview of the existing materials. It is crucial to carefully consider the makeup and characteristics of every aspect of the concrete to produce truly high-performance concrete. The study focuses on the intricate aspects of cement chemical research and distribution of particle sizes, as well as the characteristics of supplementary cementing components, understanding the molecular structure of admixtures and their potential reactions with cementing materials, as well as the physical, chemical, and mechanical qualities of aggregates, is crucial (Li & Herbert, 2012). Furthermore, it is crucial to implement rigorous quality control measures throughout every stage of the concrete production process. Collaboration between the design engineers, concrete suppliers, cement producers, and contractors is essential to guarantee the utmost quality of the result. Proper placement, finishing, and curing are crucial for ensuring the quality of the concrete, regardless of the materials used (Han et al., 2017).

5.2.1. Portland Cements

The primary ingredients for the manufacturing of Portland cement are limestone, which provides the main source of CaO, and ceramics or shales that contain specific percentages of SiO_2, Al_2O_3, and Fe_2O_3. The manufacturing technology relies on the principles of high-temperature chemical reactions. According to Taylor's research in 1997, the process of high-temperature reactions involved in the production of cement clinker can be described as follows (Shi et al., 2011):

- The clay minerals undergo decomposition at temperatures ranging from 500 to 800°C.
- Breakdown of the calcite occurs at temperatures ranging from approximately 700 to 900 degrees Celsius.
- The formation of $2CaO \cdot SiO_2$ occurs through the reactions of calcite (or lime derived from it), SiO_2, and decomposed clays at temperatures ranging from 1000 to 1300°C.
- The clinkering reactions occur at temperatures ranging from 1300 to 1450°C, resulting in the formation of $3CaO \cdot SiO_2$. A mixture of aluminate and ferrite is created to serve as a flux, aiding in the production of $3CaO \cdot SiO_2$ through the reaction of CaO and $2CaO \cdot SiO_2$ (Makul, 2020).
- As the temperature cools down to the surrounding level, the molten material undergoes a process of crystallization, resulting in the formation of ferrite and aluminate phases.
- The produced clinker is then finely ground with gypsum to regulate the setting actions of the cement. Figure 5.1 illustrates the ternary phase structure for the $CaO\text{-}Al_2O_3\text{-}SiO_2$ system, which includes the different cementitious materials (Vanikar, 2004).

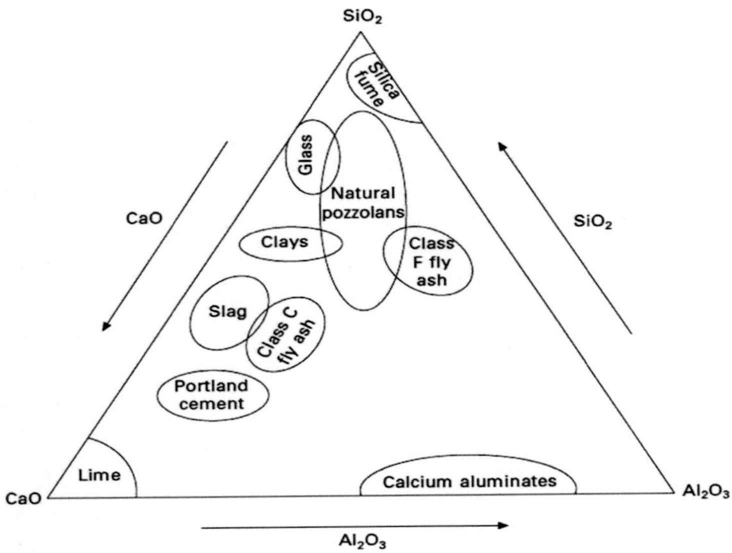

Figure 5.1. *Description of the Ternary phase diagram of the CaO--Sisystem.*

Source: Charles Ainger Creative Commons License.

Although this greatly enhances our knowledge of clinker chemical science, it still oversimplifies the actual situation by disregarding several crucial factors:

The clinker contains impurities, which make up approximately between 5 and 8% of its composition. The specific type and quantity of these impurities vary depending on the raw materials used. Various impurities, such as iron, titanium, potassium, manganese, magnesium, sulfates, alkalis, and potentially more, are present in the concrete (Wimpenny, 2009).

It should be noted that the minerals phases created are not completely pure, but rather contain additional ions, which vary depending on the specific chemical science of the raw feed. The various mineral phases are not present as individual grains; typically, every cement particle will encompass multiple phases.

These simplified versions are not crucial for regular concretes, as their physical characteristics can be fairly predicted depending on the aggregate evaluation, cement content, and water/cement ratio for both fresh and hardening states. The strength and other characteristics of concrete are not significantly affected by particulars of the cement chemistry. Still, when it comes to exceptionally high-efficiency concretes, these specifics become crucial as they typically include mineral admixtures, chemical admixtures, and especially super plasticizers (also referred to as high-range water reducers) (Kurtis, 2015). The chemical structure of these more intricate mixtures can be significantly influenced by the "minor" components that make up the cement, as well as the cement's mineralogy and composition. Finding acceptable mix designs for concretes in particular uses can be challenging due to issues with cement-super plasticizer incompatible and other adverse admixture relations. It is worth mentioning that certain types of cement used nowadays result in less long-lasting concretes compared to the older concretes produced before 1960. The increased demand for faster strength gain in the construction industry has been a major driving force. This allows for quicker form removal and ultimately leads to cost savings in construction. Through the use of finer grinding techniques and an increased tricalcium silicate content, significant progress has been made in achieving a faster rate of strength gain. However, it is important to note that these advancements also come with the drawback of greater heat of hydration and an increased risk of thermal cracking. It seems that modern concretes have a slightly higher tendency to be brittle compared to older versions (Porras et al., 2020).

5.2.2. Aggregates

Similar to types of cement, regular concrete is not greatly affected by the fundamental characteristics of the aggregates, as long as those aggregates are within the typical range of particle sizes and are not reacting chemically when mixed with cement (such as within alkali-aggregate and alkali-carbonate chemical reactions). Even aggregates that do not comply with the defined grading limits can still be used to create satisfactorily concretes, although this may result in additional expenses. However, when it comes to advanced concretes, the characteristics of the aggregate play a crucial role (Song et al., 2008). When discussing extremely strong concretes, the aggregates must possess sufficient strength and stiffness to prevent them from becoming the limiting factor in terms of strength. To achieve the desired rheological properties, it is crucial to precisely regulate the aggregate grading for the upcoming self-compacting concretes (SCC), which is going to be discussed in detail later. It might be necessary to modify the area of contact at the cement-aggregate interface, as it is often the weakest part of the system. Regrettably, the majority of research conducted in the field of concrete in recent years has primarily concentrated on the binder system. However, it is increasingly imperative to allocate equal emphasis on the aggregates that constitute a significant 70 to 80% of the concrete's volume (Ajdukiewicz & Kliszczewicz, 2007).

5.2.3. Chemical Admixtures

In the context of this discussion, the main focus is exclusively on superplasticizers, which play a crucial role in the development of advanced concretes. These kinds of additives were initially created in the 1960s. These polymers are synthetic compounds with high molecular weight. Having the ability to dissolve in water and attach themselves to the surfaces of cement particles, as well as additional cementitious particles. Their dispersing properties are highly effective, which can allow significant decreases in the quantity of water needed for sufficient workability using three mechanisms (Plank et al., 2015):

- They generate a charge that is negative upon the surfaces of the particles, which leads to repulsion between them.
- They enhance the bond between the particles or the liquid phase.
- Steric hindrance arises whenever the adsorbed polymer compounds are arranged in a manner that weakens the attraction between particles.

Currently, four primary types of super plasticizers are widely used. Melamine super plasticizers, which are condensate of formaldehyde or melamine sulfonate, were the pioneers in the field of super plasticizers and continue to be extensively utilized in the United States and Europe. And had gained a reputation for their exceptional quality and consistent performance, making them a trusted choice. There are options to choose from when it comes to naphthalene super plasticizers and are accessible in the form of sodium or calcium salts, both being polycondensate of formaldehyde along with naphthalene sulfonate. Although the sodium salt version is far more commonly utilized due to its lower cost, the calcium salt version is essential for nuclear uses with reinforced or pre-stressed concrete. This is necessary to safeguard the steel from rusting caused by chlorides. Typically, these super plasticizers are more cost-effective compared to melamine super plasticizers (Raghav et al., 2021).

Super plasticizers comprised of lignosulfonate are generally considered to be less effective compared to the other two types. As a result, become rarely employed in combination with one of the other two. In recent years, a new type of super plasticizers called polycarboxylates, also known as 'comb-type' super plasticizers, have emerged. These super plasticizers are designed to prolong the fluidity of concrete, surpassing the capabilities of naphthalene and melamine super plasticizers, without causing excessive delays in the concrete mix. Seem to be highly efficient in the production of self-compacting concretes (Nie et al., 2014).

Regrettably, the interaction between super plasticizers and cement varies due to the presence of different impurities. As a result, it is currently impossible to determine the optimal ratio between cement and super plasticizer for achieving the best outcomes. Without prior information, concrete manufacturers may need to rely on a process of experimentation to determine the optimal combination (Farzadnia et al., 2011).

5.2.4. Mineral Admixtures

Many concretes nowadays incorporate more than one mineral admixture, making it uncommon to come across concrete made solely of Portland cement. As previously stated, this holds especially true for advanced concretes, where precise control over hardly only the capacity, but also rheological characteristics and durability is crucial. Mineral admixtures also referred to as 'supplementary cementing materials,' can be either pre-blended into the cement or added independently at the batching plant, which is more prevalent in North America. Three commonly used materials in this context include fly ash, granulated blast furnace slag, and silica fume. In regular blast furnaces slag, concrete, or fly ash are commonly utilized due to their cost-effectiveness compared to Portland cement. However, silica fume, being more expensive, is not typically included in the mixture. Nevertheless, when it comes to achieving exceptional strength and minimizing permeability, incorporating silica fume is crucial. It is often used in combination with one or both the other materials. Table 5.1 provides the physical properties of several supplementary cementing materials, while Figure 5.2 displays their particle size distributions (Ahn & Kishi, 2010).

5.2.4.1. Silica Fume

Silica fume, a by-product of silicon and silicon alloy production, is frequently employed in the creation of robust concretes. It is crucial to have concretes with compression strengths exceeding 100 MPa, as this facilitates the production of high-strength concretes ranging from 60 to 100 MPa. The optimal usage is typically around 8 to 10% of the cement's weight (Panjehpour et al., 2011).

Table 5.1. *Physical Characteristics of some Supplementary Cementing Materials*

Material	**Mean size (μm)**	**Surface Area (/g)**	**Particle Shape**	**Specific Gravity**
Portland cement	10–15	<1	Angular, irregular	~3.2
Natural pozzolans	10–15	<1	Angular, irregular	Variable
Fly ash (F and C)	10–15	1–2	Mostly spherical	2.2–2.4
Silica fume	0.1–0.3	15–25	Spherical	2.2
Rice husk ash	10–20	50–100	Cellular, irregular	<2.0
Calcined clay (metakaolin)	1–2	>15	Platy	2.4

Source: Tero Kishi Creative Commons License.

Various mechanisms can impact the properties of concrete:

- It prevents the formation of calcium hydroxide on the cement-aggregate interfaces or just converts the calcium hydroxide into calcium silicate hydrate through the pozzolanic reaction, which is the reaction between silica and lime.
- This process helps to reduce the presence of large pores that the interface between the cement and aggregate, resulting in a denser material (Thomas et al., 1998).
- It alters the rheological characteristics of the newly formed concrete, resulting in a decrease in internal bleeding and a reduced need for excessive superplasticizer to attain the desired workability.
- Due to its extremely small particle size, silica fume exhibits a significant filler effect. It can effectively fill the spaces among cement grains, reducing the volume of capillary pores and expanding the overall density of the substance being filled (Das et al., 2020).

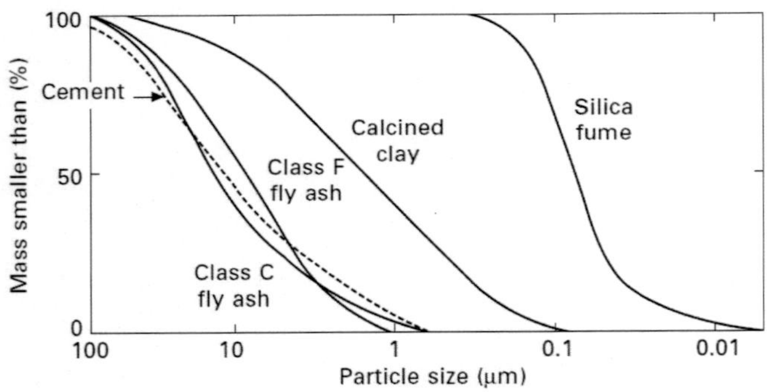

Figure 5.2. *Illustration of Particle size distributions of some supplementary cementing materials.*

Source: Richard F. Creative Commons License.

5.2.4.2. Fly Ash

Fly ash is a residue left over from the combustion of powdered coal in power plants. Concrete commonly incorporates a certain percentage of alternative materials, such as pozzolans, to reduce costs. One such material is commonly used due to its affordability and its ability to react with lime throughout cement hydration, resulting in the production of additional calcium silicate hydrate. This reaction is similar to the one that occurs with silica fume. Typically, the long-term strength of the concrete is enhanced and its durability is improved. On its own, fly ash is limited in its ability to attain concrete strengths beyond approximately 100 MPa. However, when combined with silica fume, it becomes a valuable component for achieving higher strengths (Nayak et al., 2022).

Did You Know?

Fly ash, an outcome of the combustion of coal in power stations is frequently utilized as an additional cementitious material within concrete to enhance its workability and decrease its permeability.

5.2.4.3. Blast Furnace Slag

Granulated slag from blast furnaces constitutes the residual material obtained from metallurgical procedures, which can result from the extraction of metals coming from ore or the purification of impure metals (Palod et al., 2019). They contain high levels of lime, silica, and alumina, making them somewhat comparable to Portland cement. Activated slag cement has a slow reaction rate and is commonly activated with Portland cement for commercial use. Commonly used amounts of slag in concrete mixtures range from 15 to 39% through weight of cement, although higher dosages are occasionally employed. Similar to fly ash, the use of silica fume is necessary when aiming for compressive strengths above 100 MPa (Bilim et al., 2009).

5.3. TYPES OF MODERN ADVANCED CONCRETES

Learning Objectives

- Understand the relationship between strength, porosity, and water-to-cement ratio in concrete.
- Explore the development and properties of ultra-high-strength concretes.
- Know about innovative concrete technologies for durability and sustainability.
- Discuss strategies for improving durability and sustainability in concrete technology.
- Evaluate the role of polymers and waste materials in enhancing concrete properties.

In the following sections, some of the key advanced concrete that are widely utilized in civil infrastructure projects are discussed. The selection is subjective and does not aim to encompass all types of advanced concretes. Instead, the purpose is to explain fundamental principles and showcase the various options available (Yaphary et al., 2017).

5.3.1. High Strength Concretes

In the case of most brittle materials, the strength tends to decrease as the porosity increases. One common way to describe this empirical evidence is

The strength of the material is represented by f_c, while f_{c0} refers to the strength currently zero porosity, also known as the 'intrinsic' strength. Porosity is denoted by p, and the value of k depends on the specific system. These types of expressions can also be applied to concrete, where the porosity is mainly influenced by the water-to-cement ratio (Korolev & Inozemtcev, 2013).

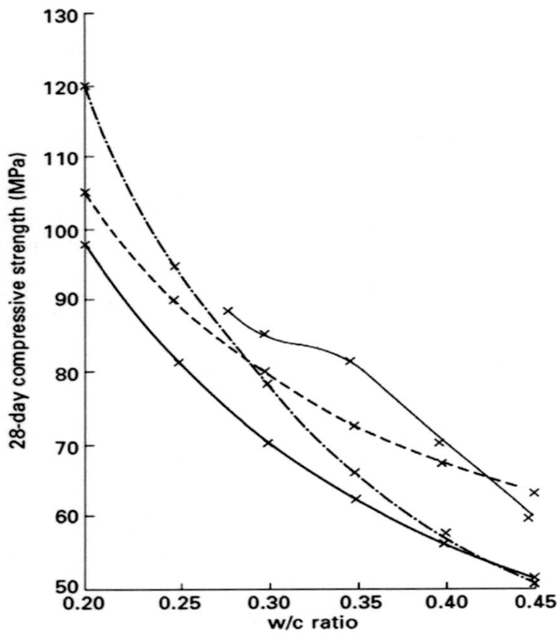

Figure 5.3. *Description of Compressive strength vs. w/c ratio for different concretes.*

Source: Suren Chen Creative Commons License.

In 1918, Duff Abrams established the w/c ratio 'law,' which asserts that the strength of concrete is solely determined by the ratio between water to cement, given the same materials. Here is an expression that represents this he values of

A along with B are determined by various factors such as the properties of the cement, the age of the concrete, and the conditions under which it was cured. This method is quite effective for regular concretes and can even be used for strengths up to approximately 120 MPa, as illustrated in Figure 5.3. Serves as the foundation for the majority of mixed design procedures (Wadekar & Pandit, 2014).

Nevertheless, when the water-to-cement ratio decreases, especially below approximately 0.3, the correlation becomes progressively less dependable. It is common to use the constant 'A' at 96.5 MPa (14,000 psi) as a reference for maximum strength in concrete for civil infrastructure. However, it is important to note that concrete can be engineered to achieve strengths far surpassing 600 MPa. Therefore, the straightforward application of the water-to-cement ratio law is ineffective when it comes to ultra-high-strength concretes. It's not surprising that Abram's law was developed before the introduction of super plasticizers and supplementary cementing products. It is evident that factors other than the w/c ratio or porosity significantly influence the strength of concretes with very low w/c ratios. Regrettably, the specific details of these extra variables are still unknown. However, they could potentially be linked to an enhanced bond between cement and aggregate, or greater bond strength found in the hydrated paste of the cement system (Shah & Ribakov, 2011).

5.3.2. Ultra-High-Strength Concretes

There are now easily accessible high-strength concrete materials within the range of about 100 to 125 MPa, at least within some regions. However, there is an additional category of concretes known as ultra-high-strength, which are also known as 'reactive powder (RPC) concretes.' These types of concretes have demonstrated impressive compressive strengths exceeding 600 MPa or flexural strengths nearing 100 MPa (Lu, 2023).

A breakthrough in construction technology emerged in France during the early 1990s. The impressive characteristics of this substance were attained through meticulous control over the concrete mixture, specifically the size of the particle distribution among all the solid materials. By optimizing the particle size distribution, a mixture can be achieved that approaches the desired density. RPC does not include any coarse aggregate. The highest aggregate size used is only 0.3 mm. This allows for the creation of a more uniform substance (Gunay et al., 2020). To enhance ductility, a small amount of steel fibers, typically as much as five percent by volume, are incorporated into the mixture. The proportions of the mix and mechanical characteristics of two distinct strength stages of RPC are provided in Table 5.2. It is important to mention that to attain higher strength levels, a certain level of thermal treatment may be necessary. A pedestrian walkway was constructed at the University of Sherbrooke, Quebec, Canada, via RPC and conventional concreting equipment and techniques. This project served as a demonstration of utilizing this unique material in real-world applications (Azmee & Shafiq, 2018).

A new commercial manufacturing process for RPC is currently being promoted as 'Ductal'. In certain cases, a heat treatment process is utilized to enhance the stability of the dimensions and durability of the material. This involves subjecting it to 90°C steam for 48 hours. When steel fibers are used, the compressive strengths typically range from 150 to 180 MPa, while the flexural strengths are around 32 MPa. The use of polypropylene fibers results in a reduction of these strengths by approximately 25%. A railroad station canopy was recently constructed in Calgary, Canada using this material (Du et al., 2021).

Another variation of this technology, which was also created in France, was recently patented and is known as CEMTEC. The multi scale approach is distinguished by significantly higher cement along with fiber contents compared to 'Ductal', while still adhering to the same underlying principles.

Table 5.2. *Typical Composition and Properties of RPC*

	RPC 200	**RPC 800**
Type V Portland cement	955 kg/	1000 kg/
Fine sand (150–400 μm)	1051 kg/	500 kg/
Silica fume	229 kg/	230 kg/m
Precipitated silica (35 /g)	10 kg/	–
Ground quartz (μm)	–	390 kg/
Superplasticizer	13 kg/	18 kg/
Steel fibers (12.5 × 180 μm)	191 kg/	–
Steel microfibers (< 3 mm long)	–	630 kg/
Total water	153 l/	180 l/
f'c (compressive strength)	170–230 MPa	490–680 MPa
E	54–60 GPa	65–75 GPa
Flexural strength	25–60 MPa	45–102 MPa

Source: Muhammad A., Creative Commons License.

The material in question can achieve impressive flexural strengths of approximately 60 MPa. Additionally, it possesses an exceptionally high impermeability. As a point of comparison, Table 5.3 provides the proportions of the mix for each of these materials (Zhang et al., 2022).

In the 1980s, Denmark made significant advancements in the development of a new family of ultra-high-strength concretes. This material is commonly known as the CRC (compact reinforced composite). The composition of this material includes a significantly minimal water/binder proportion (~0.16 or less) and incorporates steel fibers ranging from 2 to 6%. As a result, it exhibits matrix strengths ranging from 140 to 400 MPa. It sets itself apart from the materials mentioned earlier by combining them with tightly spaced conventional reinforcements made of steel. It has primarily been utilized in precast elements like staircases and balcony slabs, but it has also found application in cast-in-place scenarios (Špak et al., 2016).

There are additional types of extremely strong concrete materials that have been created, with more still being developed. These materials share several characteristics, including a significantly low water-to-binder ratio, the incorporation of silica fumes and super plasticizers, a high fiber content, restrictions on the maximum size of the aggregates, and meticulous management of particle size distribution. May also demand meticulous quality control, both during their manufacturing process and when are being installed. As a result, these materials come with a hefty price tag. Nevertheless, the exceptional durability and limitless architectural potential of concrete will continue to expand its usage in various specialized structures (Amran et al., 2022).

Table 5.3. *Compositions of Some Commercial RPCs*

Material	**Ductal® (kg/)**	**CEMTEC Multi Scale (kg/)**
Portland cement	710	1050.1
Silica fume	230	268.1
Crushed quartz	210	–
Sand	1020	514.3
Water	140	180.3
Fibers	40–160a	858b
Super plasticizer	10	44

Source: EL Reedy, Creative Commons License.

5.3.3. Fibre Reinforced Concretes

The ultra-high-strength concretes mentioned earlier are considered exceptional cases of fiber-reinforced concrete. However, their high fiber contents make them impractical for most widespread uses. Nevertheless, there is an ongoing increase in the utilization of traditional fiber-reinforced concrete (FRC) with fiber volumes below approximately 1%. This type of concrete was first created in the 1960s. It is widely recognized that simple concrete is a material that lacks flexibility, with limited ability to withstand tension and strain (Walraven, 2009). To address these drawbacks, it has become more and more prevalent to integrate small, separate fibers toward the concrete mixture. In most cases, the fibers are not included at these rates to enhance the strength of the concrete, although there may be slight improvements in strength. Instead, the primary function of the fibers seems to enhance the durability of the concrete. The purpose of these materials is to regulate the formation of cracks in the concrete and alter the properties of the material after cracking occurs. By spanning across the openings as they appear, the fibers enhance the concrete's ability to withstand further deformation. In 2001, the annual production of fiber concrete exceeded 76 million m^3, primarily used for slabs upon grade, fiber shotcrete (mainly for tunnel linings), and precast members. Various types of fibers are currently being utilized in commercial applications, with steel, polypropylene, cellulose, carbon, and glass becoming the most prevalent choices (Panzera et al., 2013).

It is important to note that, in most cases, fiber reinforcement cannot replace traditional steel reinforcement. Reinforcing bars play a crucial role in concrete by enhancing the ability to bear the weight of structural members, carrying tensile and shear stresses. On the other hand, fibers are primarily incorporated in concrete to control cracks. It is important to recognize the value of using both materials together. There are numerous applications, such as blast loading and seismic loading, where the fibers can be utilized alongside traditional reinforcement to great effect (Khalel et al., 2021).

Traditional fiber-reinforced concretes generally exhibit the stress vs. strain behavior depicted in the upper portion of Figure 5.4. Indeed, while demonstrate some ability to bear loads beyond the maximum load, and are noticeably more durable than regular concrete, and can be categorized as materials that soften under strain. Might fail due to the presence of a single significant crack. However, it is worth noting that high-performance fiber reinforced cement composites (HPFRCC) exhibit the behavior depicted in the lower section of Figure 5.4, characterized by strain-hardening and the occurrence of multiple cracks. These composites will have significantly increased fracture energies, making them highly valuable for applications involving seismic and other dynamic loading scenarios (Plizzari & Mindess, 2019).

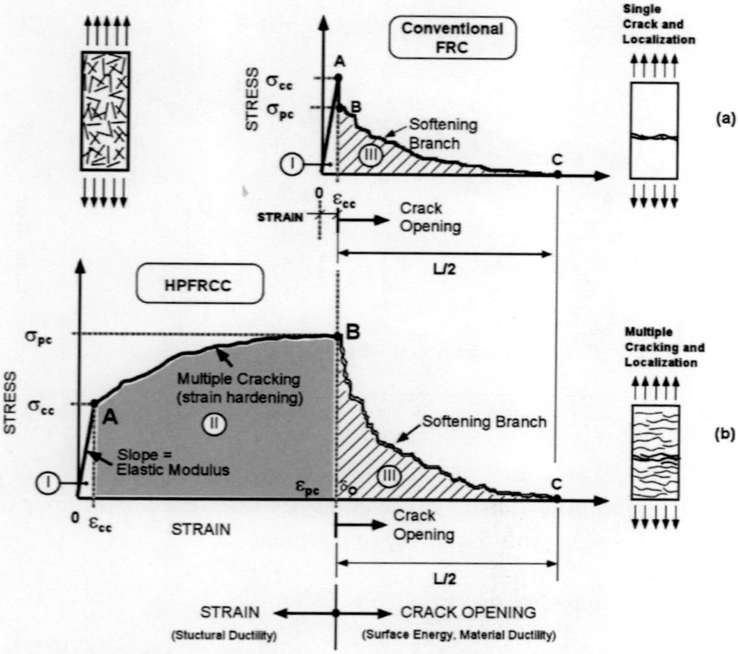

Figure 5.4. *Description of Typical stress vs. strain responses in tension of conventional FRC and high-performance fiber reinforced cement composites.*

Source: Naaman and Reinhardt Creative Commons License.

5.3.4. Self-Compacting Concrete

Self-compacting concrete also referred to as SCC, is a remarkable type of concrete that possesses the unique ability to flow effortlessly into various forms, achieving full compaction without the need for vibration. This exceptional material eliminates the risk of segregation or bleeding, making it an ideal choice for a wide range of construction projects (Goodier, 2003). This technology originated in Japan in the late 1980s and has since gained significant popularity. It is capable of producing various types of concrete, including ordinary, high-strength, and fiber-reinforced concrete. When it comes to materials, SCC may be costly compared to regular concrete. However, the benefits of easier placement and reduced labor needs make up for this expense. Additionally, it allows for precise positioning of concrete in regions with densely packed reinforcement, such as those found in seismic reinforcement or intricate structures (Hamzah et al., 2015) (Figure 5.5).

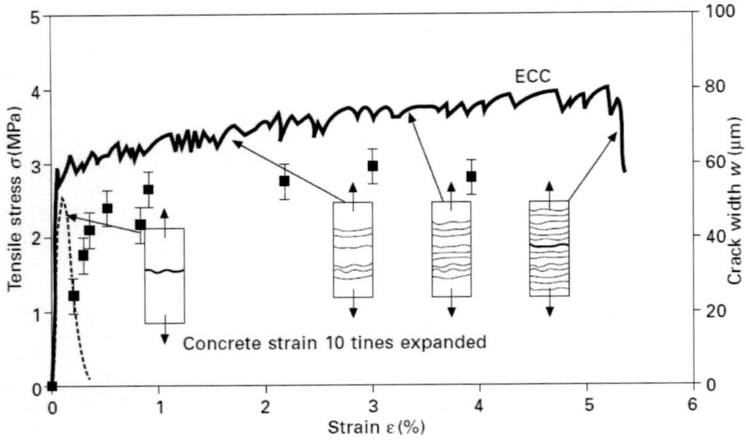

Figure 5.5. *Description of Tensile stress vs. strain response (continuous line) and crack width development (solid squares) of a typical.*

Source: Li and H. Stang Creative Commons License.

The idea behind the creation of Self-Consolidating Concrete (SCC) is quite straightforward. The fine material content within the concrete grows while keeping the quantity of mixing water relatively unchanged, in addition to the incorporation of significant quantities of super plasticizer. In SCC, the ratio for fine to rough aggregate is usually higher than the ratio of fine to rough aggregate in ordinary concrete. Specifically, a significant quantity of material alongside particle sizes smaller than 0.125 mm is necessary. This is accomplished by incorporating suitable supplementary cementing supplies, as previously explained (Okrajnov & Vasović, 2009).

The fresh self-compacting concrete (SCC) should exhibit the following characteristics:

- The filling ability of self-consolidating concrete (SCC) refers to its capacity to flow effortlessly into the structures as well as around the reinforcement bars without the need for external vibration. It ensures that the formwork is filled.
- The passing ability of the SCC refers to its capacity to flow smoothly through tight spaces in the formwork and navigate around reinforcing bars without any blockages or segregation.
- Ensuring resistance to segregation, both during its flow and once it has settled (Wani & Ganesh, 2022).

These goals can be achieved through the careful selection and proper determination of the mixture of concrete. To achieve the desired high fluidity, it is necessary to significantly restrict the amount of coarse aggregate and utilize superplasticizers. These superplasticizers not only enable a lower water-to-binder ratio but also contribute to the desired fluidity. The lower water-to-binder ratio, along with the restricted amount of coarse aggregate, results in excellent resistance to segregation. The exceptional fluidity and resistance to segregation enable the production of Self-Compacting Concrete (SCC) (Persson, 2001).

According to EFNARC (2002), the subsequent mix proportions are recommended as an initial estimate to achieve the desired fresh concrete properties:

- The water-to-powdered ratio should be between 0.8 and 1.1 by volume.
- The total powdered content should range from 400 to 600 kg/m3.
- The coarse aggregate content should be between 28-35% of the mix volume.
- The water content should be kept below 200 kg/m3.
- The remaining volume should be filled with fine aggregate (Łaźniewska-Piekarczyk, 2012).

Naturally, adjustments may be necessary to ensure that the hardened concrete meets the necessary strength and durability standards. If the initial estimate does not meet the required properties of fresh and/or hardened concrete, adjustments must be made to the mix. This can involve using alternative types or quantities of additional cementing ingredients or fine mineral fillers, employing various kinds of super plasticizers, or incorporating a viscosity-modifying agent, among other possible measures. It is important to highlight that at present, there is a lack of established or universally accepted methods for assessing the rheological characteristics of Self-Compacting Concrete (SCC). Therefore, to meet mixed design requirements or ensure quality control, various tests that are based on practical experience or empirical data are utilized, as recommended by EFNARC (Lenka & Panda, 2017).

5.3.5. High-Durability Concrete

In the field of concrete engineering, there has been a historical emphasis on strength as the main factor, with the belief that as long as the concrete is strong, it will also be able to withstand the test of time. Regrettably, this assumption is incorrect. The ideas of 'high strength' and 'high durability' are not the same; are distinct from each other. Concrete often faces significant challenges in terms of durability, which unfortunately leads to increased litigation (Swamy, 2008).

In comparison, low strength is not as much of a concern. When designing high-performance cement (or any other concrete), it is crucial to prioritize the conditions of the environment to which the finished product will be exposed. Strength considerations should follow this initial focus. Thankfully, a growing number of contemporary design codes are now mandating this approach. As an example, the current American Concrete Institute Building Code emphasizes the importance of durability demands over strength specifications (Borg et al., 2021).

Producing concrete with a small proportion of water to cementitious materials (w/cm) is crucial in protecting against various external attacks, such as industrial wastes, seawater, sulfates, and deicing salts. It serves as the initial line of defense. The decreased permeability and porosity of the concrete make it significantly more challenging for aggressive chemicals to penetrate. It is crucial to reiterate that the water-to-cement ratio, rather than the strength, plays a significant role in determining the permeability and durability of concrete. The lower water-to-cement ratio is expected to result in stronger concrete, potentially exceeding the necessary strength for structural purposes. However, it is crucial to prioritize the durability requirements above all else. Presumably, an experienced structural designer will discover ways to utilize this additional strength (Rashno & Saghaeifar, 2020).

Practice Problem

Designing a high-durability concrete mix for a coastal bridge deck requires meeting specific criteria: a compressive strength of ≥ 50 MPa at 28 days, minimum abrasion resistance of 0.10 grams per cycle, and a pore solution pH of ≥ 12.5 for steel reinforcement passivation. Select appropriate materials and formulate a concrete mix that fulfills these requirements.

Solutions to Practice Problem

To meet the requirements, it is important to use top-notch Type I Portland cement, clean and well-graded sand, as well as dense crushed stone as aggregates. The mix should be formulated with 400 kg/m³ of cement-like substances, 700 kg/m³ of fine aggregate, and 1100 kg/m³ of coarse aggregate. The water content should be adjusted to 160 kg/m³ and appropriate chemical admixtures should be added to enhance durability. Laboratory evaluations should be conducted to fine-tune the mix and ensure that it meets the desired compressive strength, corrosion resistance properties, and abrasion resistance.

Permeability is just one aspect to consider when it comes to durability. Other well-known requirements for durability must be taken into consideration, such as using sulfate-resistant cement, incorporating air infiltration for freeze-thaw durability, and opting for low alkali cement whenever the aggregates are prone to alkali-aggregate reactions. In these cases, maintaining a low water-cement ratio can help reduce the impact of potential damage (Akeed et al., 2022). It's important to note that the outer layer of concrete, known as the 'skin,' often possesses different properties compared to the rest of the concrete within a section. Due to the 'wall effect' caused by the formwork, the aggregate particles are not as tightly packed at the outermost layer of the concrete. This results in a region that tends to have a higher concentration of cement paste. Improper curing of the concrete can lead to an escalation in surface cracking caused by premature drying shrinkage. This issue becomes more pronounced as the initial water-to-cement ratio is decreased (specifically, delving into the domain of high-performance concretes) (Rex et al., 2018).

5.3.5.1. Abrasion Resistance

The abrasion obstruction of concrete plays a crucial role in various applications such as highway pavements, dam spillways, and stilling basins. In certain regions, the impact of ice abrasion is also significant. The level of abrasion resistance in concrete is closely linked to its strength. However, the properties of the coarse aggregate and the concentration of volume also play a significant role in determining durability (Bodnárová et al., 2020). Therefore, concretes with high strength, especially when silica fume is added, exhibit exceptional resistance to abrasion. There have been varying experiences when using the method of incorporating steel fibers into a low water-to-cement mix to improve abrasion resistance (Zajceva et al., 2021).

Research has shown that concrete exhibits a similar level of abrasion resistance to granite when it reaches compressive strengths that range from 120-150 MPa. Typically, when considering identical water/binder ratios, the properties of the coarse aggregate play a crucial role in determining the level of abrasion resistance (Guo et al., 2022). As an example, Laplante discovered that using granitic aggregates resulted in concrete that was more resistant to abrasion compared to dolomite or limestone aggregates. Nevertheless, a universally accepted method for accurately predicting the ability to resist abrasion of specific aggregates remains elusive. However, tests like the ones conducted in Los Angeles abrasion examination can provide some preliminary insights, particularly for initial screening (Qin et al., 2021).

5.3.5.2. Corrosion of Steel in Concrete

The rusting of steel in concrete poses a significant challenge to the durability and cost-effectiveness of our concrete infrastructure. Gaining a comprehensive understanding of the mechanisms behind the corrosion of steel in concrete over time. Whether it happens or not relies greatly on the standard of the concrete 'skin' as previously mentioned, along with the amount of thickness for the concrete cover. Various solutions to the issue of steel corrosion were suggested over time, each with its own set of benefits and limitations (Angst, 2018).

For years, galvanized reinforcing bars were utilized to combat steel corrosion. The zinc coating serves a dual purpose, acting as a protective barrier and sacrificial layer, gradually oxidizing (corroding) over time. The efficiency of the coating relies heavily on its thickness, as the duration of its effectiveness increases in direct proportion to its thickness. As an example, ASTM (1990) outlines two distinct kinds of coatings, with a thickness of approximately 85 μm and 150 μm. Thicker layers, exceeding approximately 200 μm, have the potential to weaken the bond among the concrete along with the rebar (Rajbanul Akhond et al., 2024). Although galvanizing proves to be effective in preventing corrosion caused by carbonizing the concrete covering, its effectiveness seems to be reduced when corrosion is caused by chlorides. In addition, if the coating is not thick enough, it may easily break when the rebar is being bent or dealt with at the job site. This can result in rapid localized corrosion. Unfortunately, welding is not possible with galvanized steel. However, even though galvanized rebar is more expensive, it appears to have a strong demand in the market.

The epoxy coating on reinforcing bars serves as a protective barrier, preventing the steel from being exposed to harsh environmental conditions. In general, the depth of the coating made of epoxy is usually specified to fall within the limits of 130 μm to 300 μm, as per ASTM 1993. Additionally, there are specific guidelines regarding the maximum amount of defects allowed in the coating. The epoxy should have sufficient flexibility to allow for the bending of the bar without causing any damage. If the coating is damaged, it can result in significant localized corrosion and ultimately cause the rebar to fail. In certain warm marine environments, there have been cases where the epoxy coating has detached from the steel. Aside from the expensive price tag, one significant drawback of epoxy coating is the weakened bond between the concrete along the steel. This can lead to an increased risk of cracking and necessitate the application of larger anchorage or lap lengths. In certain situations, stainless steel reinforcement bars may be utilized in conditions that involve extreme exposure. Although their effectiveness in preventing corrosion is undeniable, the high cost associated with them greatly restricts their usage (Page, 2007).

Corrosion inhibitors of different varieties are now becoming more prevalent. These substances do not function as obstacles to harmful substances, but rather as compounds that minimize the steel's susceptibility to corrosion. There are two different categories of corrosion inhibitors: Passivating film upon the steel surface is stabilized and reinforced by anodic inhibitors within the high pH surroundings of concrete. Cathodic inhibitors are adsorbed onto the steel surface, where they act as a barrier to the reduction of oxygen (which is the principal cathodic reaction for steel in concrete). However, it must be remembered that corrosion inhibitors are effective only if used in otherwise good concrete; they are not a panacea for corrosion in poorly designed or placed concrete (Veleva et al., 1998).

The most common anodic corrosion inhibitor is currently calcium nitrite, which acts essentially by increasing the level of chloride necessary to initiate corrosion. Cathodic inhibitors are absorbed on the steel surface, creating a protective barrier against the oxygen reduction process, which is the main cathodic response for steel in concrete. It is important to note that inhibitors of corrosion are only effective when used in well-made concrete. And cannot solve the problem of corrosion in inadequately constructed or improperly placed concrete (Mehta & Gerwick, 1996).

5.3.6. Polymer Modified Concretes

As mentioned previously, plain concrete is a brittle material. Therefore, by incorporating polymers, which possess ductility and strength, it is possible to create a composite material that exhibits both strength and toughness. In addition, the inclusion of polymers can help decrease permeability, thereby enhancing the longevity of concrete (Aziz et al., 2023). However, implementing this approach proves to be challenging and costly in practice, with expenses ranging from three to six times that of plain concrete. As a result, the utilization of polymer-cement composites remains limited. However, with the growing emphasis on durability and repair factors, these materials are expected to gain a larger market share (Liu et al., 2019).

5.3.6.1. Latex-modified Concrete

One commonly used method to incorporate a polymer into concrete is by adding polymer latex to the mixture over the mixing process. Introducing a game-changer in the world of civil infrastructure: latex-modified concrete or LMC. Some commonly used polymers for this application include vinylidine chloride, vinyl acetate, styrene-butadiene, and acrylic esters. The resulting materials have a primary application in bonding agents, overlays, and patching (Shete & Upase, 2014).

Figure 5.6 provides a schematic representation of the formation process of the polymer film in LMC. As the water evaporates the latex particles come together and create a smooth film as the cement reacts and removes more water. (Please keep in mind that instead of prolonged moist curing, it is advisable to let the concrete dry out following a day or two to facilitate the creation of a protective film). The polymer film develops within the capillary holes and at the interface between the paste of cement and aggregate, thereby strengthening the bond between the paste and aggregate, and restricting the ingress of water via the capillary pore network. The films formed have considerable strength and resilience, thereby enhancing the resistance to crack propagation within the matrix (Sreekala et al., 2009).

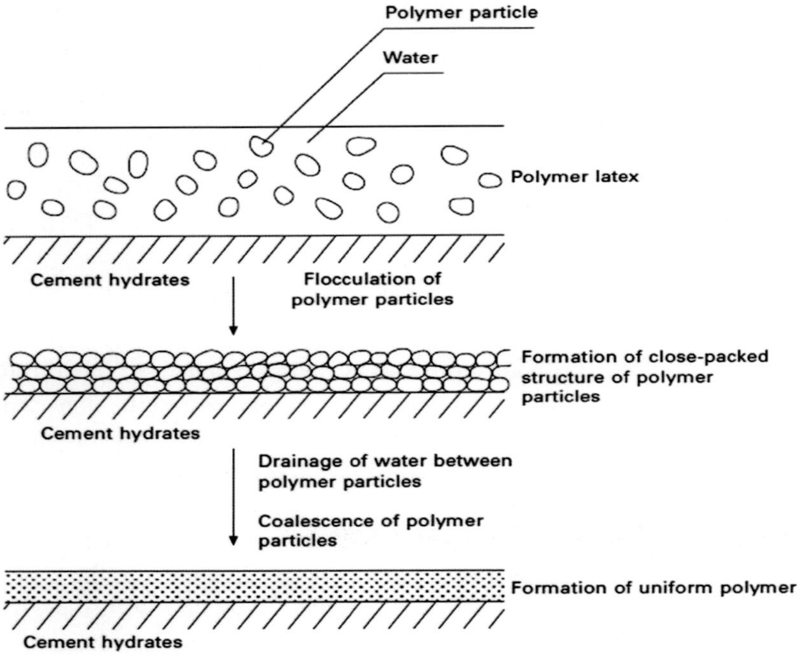

Figure 5.6. *Schematic representation of polymer film formation in LMC.*

Source: George Voy Creative Commons License.

LMC exhibits superior strength compared to unmodified concrete when subjected to the same curing conditions, especially in compression, tension, and flexure. LMC exhibits a reduced modulus of elasticity compared to plain concrete, yet it demonstrates greater strain at failure. Considering its primary use as a repair overlay, it is important to note that LMC exhibits superior bonding properties to both existing concrete and reinforcing bars. However, its strength may diminish over time when exposed to water for extended periods. LMC significantly surpasses plain concrete in terms of durability for various reasons (Talib, 2018). The polymer film significantly decreases the ability to flow the concrete. The enhanced ability to resist tensile cracking results in a decrease in the occurrence of cracks that can allow water to penetrate the material. Additionally, it aids in narrowing the cracks by forming bridges across them, similar to the function of fibers. Ultimately, the reduced water-to-cement ratio employed in LMC concretes contributes to enhanced durability.

In recent times, there has been a growing interest in the integration of fibers with LMC. Research carried out by Xu et al., (2004a, 2004b) has demonstrated the superior strength and toughness of these materials compared to LMC or FRC individually. These findings suggest that these materials could be effectively utilized for repair purposes and in situations that require high-toughness materials (Bothra & Ghugal, 2015).

5.3.6.2. Polymer-impregnated Concrete

Polymer-impregnated concrete (PIC) is created by saturating the concrete with a liquid monomer, which is then transformed into a solid polymer within the pores of the concrete. There are two methods to initiate the polymerization process (Czarnecki, 2006): one is through the application of gamma radiation, and the other involves the application of a substance called a catalyst and heat. Some commonly used monomers include methyl methacrylate (MMA), and styrene, along with a monomer of plexiglass. These monomers have low viscosities, making them suitable for penetrating the concrete pore system (Yehia & Falah, 2020).

Concrete used in civil infrastructure can exhibit significantly higher strength compared to plain concrete, with a potential increase of two to four times. However, it is important to note that this enhanced strength comes at the expense of increased brittleness. Similar to LMC, polymer impregnation enhances the strength of both the hydrated paste of cement along the cement-aggregate interface (Ohama, 1997). What's even more significant is that it highly enhances the longevity of the concrete by effectively filling all the pores with the polymer, thus preventing the penetration of harmful chemicals. Concrete for civil infrastructure is a material that comes with a high price tag, resulting in limited utilization in practical applications. Once more, it is primarily utilized in situations where a prolonged period of maintenance-free operation is sought after, or in cases where carrying out repairs would be challenging (Maiti & Kirtania, 1986).

5.3.7. 'Green' Concrete

The concrete industry has a significant environmental impact due to its extensive use of resources from nature, making it the largest consumer of these resources globally. Every type of Portland cement necessitates approximately 1.5 tons of essential components for its manufacturing process and consumes a significant amount of energy ranging from 4000 to 7500 megajoules. The production of Portland cement is a significant contributor to greenhouse gas emissions, with each ton of cement produced resulting in the release of approximately one ton of CO_2 into the atmosphere. The cement industry currently contributes to approximately 7% of global CO_2 pollutants. For the cement sector to maintain its sustainability, it is necessary to implement substantial changes in the methods of specifying, producing, using, and recycling concrete (Imbabi et al., 2012).

Did You Know?

Environmentally friendly concrete known as "Green" concrete formulations aims to minimize its environmental impact by incorporating recycled materials, low-carbon cement alternatives, and industrial by-products.

An effective solution to this issue involves increasing the percentage of flying ash (or granulated furnace slag) in concrete. This approach is beneficial from the economic and environmental perspectives. Currently, there is significant progress in the development of a type of concrete known as 'high-performance, high-volume fly ash concrete.' These types of concretes can be described as (Osial et al., 2022):

- Having a minimum of 50% fly ash about the weight of the cementing materials.
- With a Portland cement content below 200 kg/m3.
- With a water content below 130 kg/m3.
- With a water/cementing content ratio lower than 0.35.

Due to the higher pozzolanic content, the strength of these concretes takes longer to fully develop compared to conventional concretes. This slower strength gain should be factored into the construction timeline. In contrast, the strength of these concretes at 91 days and 365 days maybe 20% and 40% higher, respectively, than the strength at 28 days. This offers a significant advantage in terms of the concrete's durability (Suhendro, 2014).

It seems that there is no significant impact on whether fly ash is included at the mixing facility or pre-blended into the cement. The achievable maximum strength of these concretes is still uncertain, but it has been proven that compressive strengths exceeding 40 MPa can be achieved within 28 days.

High-volume fly ash concretes, when produced correctly, exhibit exceptional durability due to their low permeability. Additionally, they exhibit a higher resistance to cracking compared to conventional concretes with similar strength. One possible explanation for this can be found by examining the two standard mixes presented in Table 5.4 (Provis, 2014). The large-scale fly ash mix exhibits a reduced water-to-cement ratio and a significantly decreased paste volume (23.4% compared to 29.5%), while both mixes maintain a similar slump of 125–150 mm. Due to these factors, the fly ash concrete experiences a significant reduction in drying shrinkage and heat of hydration, resulting in a decreased risk of cracking. It is important to highlight that, for high-volume fly ash mixes, proper curing is crucial, even more so than for regular concretes. For optimal results, it is advised to provide a minimum of seven days for wet curing at temperatures above 10°C for large quantities of fly ash concrete (Golewski, 2021).

Table 5.4. *Comparison of Mix Proportions for 25 MPa Concrete*

	Conventional Concrete		**HVFA Concrete**	
	By Mass (Kg/)	**By Volume ()**	**By Mass (Kg/)**	**By Volume ()**
Cement	307	0.097	154	0.049
Fly ash	–	–	154	0.063
Water	178	0.178	120	0.120
Entrapped air (2%)	–	0.020	–	0.020
Coarse aggregate	1040	0.387	1210	0.449
Fine aggregate	825	0.318	775	0.299
Total	2350	1.000	2413	1.000
w/cm	0.58	–	0.38	–

Source: Wood Head, Creative Commons License.

Although blast furnace slag and fly ash are currently receiving significant attention as potential cement substitutes, there are additional waste substances that can be integrated into concrete. These are commonly used as substitutes for traditional natural aggregates. (It is important to note that aggregates constitute a significant portion, ranging from 70 to 80%, of the overall volume of concrete.) There are numerous options for obtaining these potential aggregates, including mineral wastes, incinerator residues, slags, and recycled concrete. Nevertheless, it is crucial to bear in mind three key factors when assessing these waste materials (Luhar et al., 2021):

- *Economy:* What is the quantity of material that is accessible, the distance it needs to be transported, the level of beneficiation it will need, and how will it impact the fundamental concrete mix design?
- Will it have any negative reactions with other materials in the mix, such as cement or admixtures?
- Will the material have any impact on the strength, elastic modulus, and volume security (shrinkage, creep) of the concrete (Panda et al., 2020)?

SUMMARY

- This chapter, investigates the world of contemporary advanced concrete, delving into its structure, materials, and different types. The text starts by introducing modern advanced concrete and proceeds to explore the main components utilized, which encompass chemical admixtures, Portland cement, aggregates, and mineral admixtures like silica fume, fly ash, and blast furnace slag.
- The chapter further explores different types of modern advanced concretes, such as high-strength concretes, ultra-high-strength concretes, fiber-reinforced concretes, self-compacting concrete, high-durability concrete, and polymer-modified concretes. Each type is examined in terms of its properties, applications, and benefits in civil infrastructure projects.
- Special attention is given to high-durability concrete, highlighting its resistance against abrasion and corrosion, crucial for structures exposed to harsh environments. Additionally, polymer-modified concretes, including latex-modified concrete and polymer-impregnated concrete, are discussed for their unique characteristics and applications.
- The chapter concludes with a discussion on 'green' concrete, emphasizing its low environmental impact and its potential to contribute to sustainable construction practices. Overall, this chapter provides a comprehensive overview of advanced concrete technologies essential for modern civil infrastructure projects.

MULTIPLE CHOICE QUESTIONS

1. What is one of the defining characteristics of modern advanced concrete?

a. Low strength

b. Traditional materials

c. Innovative additives

d. Limited durability

2. Which of the following is NOT a material commonly used in modern advanced concrete?

a. Portland cement

b. Sand

c. Fly ash

d. Silica fume

3. What is the primary purpose of chemical admixtures in concrete?

a. Increase aggregate size

b. Decrease workability

c. Enhance strength and durability

d. Reduce water content

4. Which mineral admixture is known for its pozzolanic properties and is a byproduct of coal combustion?

a. Silica fume

b. Fly ash

c. Blast furnace slag

d. Latex

5. Which type of concrete is specifically engineered to resist abrasion and chemical attack?

a. High strength concrete

b. Ultra-high-strength concrete

c. High-durability concrete

d. Self-compacting concrete

6. What is a characteristic feature of 'green' concrete?

a. Low environmental impact

b. High carbon emissions

c. Rapid degradation

d. Excessive water usage

REVIEW QUESTIONS

1. Discuss the significance of innovative additives in modern advanced concrete. How do they contribute to improving concrete properties?
2. Compare and contrast the characteristics of high-strength concrete and ultra-high-strength concrete. What are the key differences in their compositions and applications?
3. Explain the role of chemical admixtures in concrete mixtures. Provide examples of common chemical admixtures and their functions.

4. Evaluate the benefits of using mineral admixtures such as fly ash, silica fume, and blast furnace slag in concrete production. How do they enhance the performance of concrete?
5. Describe the properties of high-durability concrete and its applications in civil infrastructure. Discuss the factors contributing to its resistance against abrasion and corrosion.
6. Analyze the advantages and challenges associated with polymer-modified concretes. How do latex-modified concrete and polymer-impregnated concrete differ in their compositions and performance?
7. Assess the environmental impact of 'green' concrete compared to traditional concrete mixtures. What measures are taken to reduce the carbon footprint of concrete production, and what are the potential benefits of sustainable construction practices?

Answers to Multiple Questions

1. (c) 2. (b) 3. (c) 4. (b) 5. (c) 6. (a)

REFERENCES

1. Ahn, T. H., & Kishi, T. (2010). Crack self-healing behavior of cementitious composites incorporating various mineral admixtures. *Journal of Advanced Concrete Technology, 8*(2), 171–186. https://doi.org/10.3151/jact.8.171.
2. Ajdukiewicz, A. B., & Kliszczewicz, A. T. (2007). Comparative tests of beams and columns made of recycled aggregate concrete and natural aggregate concrete. *Journal of Advanced Concrete Technology, 5*(2), 259–273. https://doi.org/10.3151/jact.5.259.
3. Angst, U. M. (2018). Challenges and opportunities in corrosion of steel in concrete. *Materials and Structures, 51*(1), 4. https://doi.org/10.1617/s11527-018-1186-0.
4. Awoyera, P. O., Oluyemi, F., Bheel, N., Aluko, O., de Azevedo, A. R., & Olalusi, O. B. (2024). Morphology and strength of concrete incorporating natural rubber latex and plastic fibers. In *Green Materials in Civil Engineering* (Vol. 3, pp. 201–217). Woodhead Publishing.
5. Aziz, I. A., Mohammad, K. I., & Al-Luhybi, A. S. (2023). Mechanical properties and strength development of polymer modified concrete: A contribution study. In *AIP Conference Proceedings* (Vol. 2651, pp. 3698–3699). AIP Publishing. https://doi.org/10.1063/5.0122432.

6. Azmee, N. M., & Shafiq, N. (2018). Ultra-high-performance concrete: From fundamental to applications. *Case Studies in Construction Materials, 9*, 197. https://doi.org/10.1016/j.cscm.2018.02.007.

7. Bautista-Gutierrez, K. P., Herrera-May, A. L., Santamaría-López, J. M., Honorato-Moreno, A., & Zamora-Castro, S. A. (2019). Recent progress in nanomaterials for modern concrete infrastructure: Advantages and challenges. *Materials, 12*(21), 3548. https://doi.org/10.3390/ma12213548.

8. Bilim, C., Atiş, C. D., Tanyildizi, H., & Karahan, O. (2009). Predicting the compressive strength of ground granulated blast furnace slag concrete using artificial neural network. *Advances in Engineering Software, 40*(5), 334–340. https://doi.org/10.1016/j.advengsoft.2008.10.002.

9. Bodnárová, L., Ťažký, M., Ťažká, L., Hela, R., Pikna, O., & Sitek, L. (2020). Abrasive wear resistance of concrete in connection with the use of crushed and mined aggregate, active and non-active mineral additives, and the use of fibers in concrete. *Sustainability, 12*(23), 9920. https://doi.org/10.3390/su12239920.

10. Borg, R. P., Cuenca, E., Garofalo, R., Schillani, F., Nasner, M. L., & Ferrara, L. (2021). Performance assessment of ultra-high durability concrete produced from recycled ultra-high durability concrete. *Frontiers in Built Environment, 7*, 648220. https://doi.org/10.3389/fbuil.2021.648220.

11. Bothra, S. R., & Ghugal, Y. M. (2015). Polymer-modified concrete. *International Journal of Research in Engineering and Technology, 4*, 845–848. https://doi.org/10.15623/ijret.2015.0406041.

12. Czarnecki, L. (2006). Essential concrete-polymer composites. In *International Symposium Polymers in Concrete* (Vol. 2, pp. 13–20). Oficinas Gráficas de Barbosa & Xavier, Lda.

13. Das, S. K., Singh, S. K., Mishra, J., & Mustakim, S. M. (2020). Effect of rice husk ash and silica fume as strength-enhancing materials on properties of modern concrete—A comprehensive review. In *Emerging Trends in Civil Engineering: Select Proceedings of ICETCE 2018* (pp. 253–266). https://doi.org/10.1007/978-981-15-3360-5_24.

14. Du, J., Meng, W., Khayat, K. H., Bao, Y., Guo, P., Lyu, Z., & Wang, H. (2021). New development of ultra-high-performance concrete (UHPC). *Composites Part B: Engineering, 224*, 109220. https://doi.org/10.1016/j.compositesb.2021.109220.

15. Farzadnia, N., Ali, A. A. A., & Demirboga, R. (2011). Incorporation of mineral admixtures in sustainable high-performance concrete. *International Journal of Sustainable Construction Engineering and Technology, 2*(1), 9874–9876.

16. Feizbahr, M., & Pourzanjani, P. (2024). The impact of advanced concrete technologies on modern construction and aesthetics. *Journal of Review in Science and Engineering, 2024*, 1–9.

17. Garden, H. N. (2004). Use of advanced composites in civil engineering infrastructure. *Proceedings of the Institution of Civil Engineers-Structures and Buildings, 157*(6), 357–368. https://doi.org/10.1680/stbu.2004.157.6.357.

18. Golewski, G. L. (2021). Green concrete based on quaternary binders with significant reduction of CO_2 emissions. *Energies, 14*(15), 4558. https://doi.org/10.3390/en14154558.

19. Goodier, C. I. (2003). Development of self-compacting concrete. *Proceedings of the Institution of Civil Engineers-Structures and Buildings, 156*(4), 405–414. https://doi.org/10.1680/stbu.2003.156.4.405.

20. Gunay, A. R., Karadeniz, S., & Kaya, M. (2020). An experimental study on the dynamic behavior of ultra-high-strength concrete. *Applied Sciences, 10*(12), 4170. https://doi.org/10.3390/app10124170.

21. Guo, J., Gao, M., Wang, K., & Zhang, P. (2022). Mechanisms and influential variables on the abrasion resistance of hydraulic concrete. *Nanotechnology Reviews, 11*(1), 2997–3019. https://doi.org/10.1515/nanorev-2022-0012.

22. Hamzah, A. F., Wan Ibrahim, M. H., Jamaluddin, N., Jaya, R. P., & Zainal Abidin, N. E. (2015). Cementitious materials usage in self-compacting concrete: A review. *Advanced Materials Research, 1113*, 153–160. https://doi.org/10.4028/www.scientific.net/AMR.1113.153.

23. Han, B., Zhang, L., & Ou, J. (2017). *Smart and Multifunctional Concrete Toward Sustainable Infrastructures* (Vol. 2, pp. 369–377). Springer.

24. Hollaway, L. C. (2003). The evolution of and the way forward for advanced polymer composites in the civil infrastructure. *Construction and Building Materials, 17*(6–7), 365–378. https://doi.org/10.1016/S0950-0618(03)00009-4.

25. Imbabi, M. S., Carrigan, C., & McKenna, S. (2012). Trends and developments in green cement and concrete technology. *International Journal of Sustainable Built Environment, 1*(2), 194–216. https://doi.org/10.1016/j.ijsbe.2012.09.001.

26. Karbhari, V. M., & Seible, F. (2000). Fiber reinforced composites–Advanced materials for the renewal of civil infrastructure. *Applied Composite Materials, 7*, 95–124. https://doi.org/10.1023/A:1008314622844.

27. Khalel, H., Khan, M., Starr, A., Khan, K. A., & Muhammad, A. (2021).

Performance of engineered fiber reinforced concrete (EFRC) under different load regimes: A review. *Construction and Building Materials, 306*, 124692. https://doi.org/10.1016/j.conbuildmat.2021.124692.

28. Kioumarsi, M., & Plevris, V. (2024). Advanced concrete and construction materials for sustainable structures. *Sustainability, 16*(4), 1427. https://doi.org/10.3390/su16041427.
29. Korolev, E. V., & Inozemtcev, A. S. (2013). Preparation and research of high-strength lightweight concrete based on hollow microspheres. *Advanced Materials Research, 746*, 285–288. https://doi.org/10.4028/www.scientific.net/AMR.746.285.
30. Kurtis, K. E. (2015). Innovations in cement-based materials: Addressing sustainability in structural and infrastructure applications. *MRS Bulletin, 40*(12), 1102–1109. https://doi.org/10.1557/mrs.2015.273.
31. Łaźniewska-Piekarczyk, B. (2012). The influence of selected new generation admixtures on the workability, air-voids parameters and frost-resistance of self-compacting concrete. *Construction and Building Materials, 31*, 310–319. https://doi.org/10.1016/j.conbuildmat.2011.12.026.
32. Lenka, S., & Panda, K. C. (2017). Effect of metakaolin on the properties of conventional and self-compacting concrete. *Advances in Concrete Construction, 5*(1), 031. https://doi.org/10.12989/acc.2017.5.1.031.
33. Li, V. C., & Herbert, E. (2012). Robust self-healing concrete for sustainable infrastructure. *Journal of Advanced Concrete Technology, 10*(6), 207–218. https://doi.org/10.3151/jact.10.207.
34. Liu, J., Yu, C., Shu, X., Ran, Q., & Yang, Y. (2019). Recent advances of chemical admixtures in concrete. *Cement and Concrete Research, 124*, 105834. https://doi.org/10.1016/j.cemconres.2019.105834.
35. Lu, J. X. (2023). Recent advances in high strength lightweight concrete: From development strategies to practical applications. *Construction and Building Materials, 400*, 132905. https://doi.org/10.1016/j.conbuildmat.2023.132905.
36. Mehta, P. K., & Gerwick, B. C. (1996). Concrete in service of modern world. In *Proc. Int. Conf. on Concrete in Service of Mankind* (Vol. 3, pp. 1–28). University of Dundee.
37. Nayak, D. K., Abhilash, P. P., Singh, R., Kumar, R., & Kumar, V. (2022). Fly ash for sustainable construction: A review of fly ash concrete and its beneficial use case studies. *Cleaner Materials, 6*, 100143. https://doi.org/10.1016/j.clema.2022.100143.
38. Ohama, Y. (1997). Recent progress in concrete-polymer composites.

Advanced Cement Based Materials, 5(2), 31–40. https://doi.org/10.1016/S0958-9465(97)00007-2.

39. Okrajnov, B. R., & Vasović, D. (2009). Self-compacting concrete and its application in contemporary architectural practice. *Spatium,* (20), 28–34. https://doi.org/10.2298/SPAT0920028O.
40. Osial, M., Pregowska, A., Wilczewski, S., Urbańska, W., & Giersig, M. (2022). Waste management for green concrete solutions: A concise critical review. *Recycling, 7*(3), 37. https://doi.org/10.3390/recycling7030037.
41. Page, C. L. (2007). Corrosion and protection of reinforcing steel in concrete. *Durability of Concrete and Cement Composites, 7*(4), 136–186. https://doi.org/10.1016/S0958-9465(97)00007-2.
42. Palod, R., Deo, S.V., & Ramtekkar, G.D. (2019). Utilization of waste from steel and iron industry as replacement of cement in mortars. Journal of Material Cycles and Waste Management, 21(6), 1361-1375. http://dx.doi.org/10.1007/s10163-019-00889-3.
43. Panda, R., Sagarika, P., & Mohanty, A. M. (2020). Geopolymer concrete: An emerging green material for modern construction. *International Journal of Engineering and Advanced Technology, 9*, 4046–4050. https://doi.org/10.35940/ijeat.A7804.109119.
44. Persson, B. (2001). A comparison between mechanical properties of self-compacting concrete and the corresponding properties of normal concrete. *Cement and Concrete Research, 31*(2), 193–198. https://doi.org/10.1016/S0008-8846(00)00443-4.
45. Plizzari, G., & Mindess, S. (2019). Fiber-reinforced concrete. In *Developments in the Formulation and Reinforcement of Concrete* (Vol. 3, pp. 257–287). Woodhead Publishing. https://doi.org/10.1016/B978-0-08-100454-6.00009-1.
46. Porras, Y., Jones, C., & Schmiedeke, N. (2020). Freezing and thawing durability of high early strength portland cement concrete. *Journal of Materials in Civil Engineering, 32*(5), 04020077. https://doi.org/10.1061/(ASCE)MT.1943-5533.0003087.
47. Provis, J. L. (2014). Green concrete or red herring? –Future of alkali-activated materials. *Advances in Applied Ceramics, 113*(8), 472–477. https://doi.org/10.1179/1743676114Y.0000000162.
48. Qin, Q., Meng, Q., Yang, H., & Wu, W. (2021). Study of the anti-abrasion performance and mechanism of coral reef sand concrete. *Construction and Building Materials, 291*, 123263. https://doi.org/10.1016/j.conbuildmat.2021.123263.

49. Raghav, M., Park, T., Yang, H. M., Lee, S. Y., Karthick, S., & Lee, H. S. (2021). Review of the effects of supplementary cementitious materials and chemical additives on the physical, mechanical, and durability properties of hydraulic concrete. *Materials, 14*(23), 7270. https://doi.org/10.3390/ma14237270.
50. Rashno, A., & Saghaeifar, A. (2020). Durability of ultra-high-performance self-compacting concrete with hybrid fibers. *Emerging Materials Research, 9*(2), 331–341. https://doi.org/10.1680/jemmr.19.00032.
51. Rex, J., Babu, J. S., & Reddy, S. P. S. (2018). Strength and durability aspects of bacterial concrete. *International Journal of Innovative Technology and Exploring Engineering, 8*(22), 9–13. https://doi.org/10.35940/ijitee.L2342.1081218.
52. Shah, A. A., & Ribakov, Y. (2011). Recent trends in steel fibered high-strength concrete. *Materials & Design, 32*(8–9), 4122–4151. https://doi.org/10.1016/j.matdes.2011.03.021.
53. Shete, G. N., & Upase, K. S. (2014). Evaluation of compressive strength and water absorption of styrene butadiene rubber (SBR) latex modified concrete. *International Journal of Modern Engineering Research (IJMER), 4*(10), 40–44. https://doi.org/10.9790/1684-0410040404.
54. Shi, C., Jiménez, A. F., & Palomo, A. (2011). New cements for the 21st century: The pursuit of an alternative to Portland cement. *Cement and Concrete Research, 41*(7), 750–763. https://doi.org/10.1016/j.cemconres.2011.03.016.
55. Song, G., Gu, H., & Mo, Y. L. (2008). Smart aggregates: Multi-functional sensors for concrete structures—a tutorial and a review. *Smart Materials and Structures, 17*(3), 33001. https://doi.org/10.1088/0964-1726/17/3/033001.
56. Špak, M., Kozlovská, M., Struková, Z., & Bašková, R. (2016). Comparison of conventional and advanced concrete technologies in terms of construction efficiency. *Advances in Materials Science and Engineering, 2016*, 654–660. https://doi.org/10.1155/2016/9872621.
57. Sreekala, R., Muthumani, K., Sathish Kumar, K., & Gopalakrishnan, N. (2009). Natural rubber latex modified fibre concrete—Initial studies for structural performance. *Indian Concrete Journal, 83*(10), 29–34.
58. Suhendro, B. (2014). Toward green concrete for better sustainable environment. *Procedia Engineering, 95*, 305–320. https://doi.org/10.1016/j.proeng.2014.12.149.
59. Veleva, L., Castro, P., Hernandez-Duque, G., & Schorr, M. (1998). The corrosion performance of steel and reinforced concrete in a tropical

humid climate: A review. *Corrosion Reviews, 16*(3), 235–284. https://doi.org/10.1515/CORRREV.1998.16.3.235.

60. Wadekar, A. P., & Pandit, R. D. (2014). Study of different types fibers used in high strength fiber reinforced concrete. *International Journal of Innovative Research and Advanced Engineering, 1*, 225–230.
61. Walraven, J. C. (2009). High performance fiber reinforced concrete: Progress in knowledge and design codes. *Materials and Structures, 42*(9), 1247–1260. https://doi.org/10.1617/s11527-009-9510-7.
62. Wani, T. A., & Ganesh, S. (2022). Study on fresh properties, mechanical properties and microstructure behavior of fiber reinforced self-compacting concrete: A review. *Materials Today: Proceedings, 62*, 6663–6670. https://doi.org/10.1016/j.matpr.2022.01.278.
63. Wimpenny, D. (2009). Low carbon concrete-options for the next generation of infrastructure. *Concrete Solutions, 9*, 41–41.
64. Yaphary, Y. L., Lam, R. H., & Lau, D. (2017). Chemical technologies for modern concrete production. *Procedia Engineering, 172*, 1270–1277. https://doi.org/10.1016/j.proeng.2017.02.172.
65. Yehia, S., & Falah, M. W. (2020). Study on fresh and hardened characteristics of polymer impregnated no-fines concrete. *International Journal of Advanced Science and Technology, 29*(3), 13472–13483.

CHAPTER 6

ADVANCED CEMENTITIOUS COMPOSITES

LEARNING OBJECTIVES

At the end of this chapter, readers will be able to:

1. Understand the evolution and significance of fiber-reinforced cementitious composites in modern construction.
2. Understand the key factors influencing mechanical properties and performance of fiber-reinforced cementitious composites.
3. Know the applications of fiber type, volume fraction, matrix variation, and processing methods in enhancing the properties of cement-based composites.
4. Understand the relationship between fiber-cement bond properties and the mechanical behavior of fiber-reinforced cementitious composites.
5. Understand the key factors in producing high-strength concrete.
6. Comprehend the characteristics and benefits of micro silica (MS) concrete.
7. Recognize the unique properties and applications of densified with small particles (DSP) materials.

INTRODUCTORY EXAMPLE

TechStructures Inc specializes in advanced construction solutions has recently completed the construction of a state-of-the-art bridge using their proprietary Fiber-Reinforced Cementitious Composite (FRCC). This bridge, spanning a river in a densely populated urban area, faced numerous challenges such as high traffic volume, corrosive environment due to exposure to water and chemicals, and the need for rapid construction to minimize disruption to commuters.

To address these challenges, TechStructures Inc. utilized their advanced FRCC, which incorporates high-strength glass fibers into the cement matrix. These fibers enhance the structural integrity and durability of the composite, making it highly resistant to corrosion, fatigue, and impact.

During construction, the use of FRCC facilitated rapid installation, reducing construction time by 30% compared to conventional methods. Despite its lightweight nature, the FRCC bridge boasts exceptional strength, allowing it to accommodate heavy traffic loads while maintaining structural integrity over its projected lifespan of 100 years.

This project exemplifies the transformative potential of advanced cementitious composites in revolutionizing infrastructure development, offering enhanced performance, durability, and sustainability for the built environment. TechStructures Inc. continues to lead the way in advancing the field of cementitious materials, driving innovation and shaping the future of construction worldwide.

UNIT INTRODUCTION

The chapter regarding Advanced Cementitious Composites introduces a range of innovative materials and techniques that have the potential to revolutionize the field of modern construction. The field of cementitious composites has witnessed a remarkable transformation, surpassing conventional concrete to provide unparalleled durability, reliability, and sustainability, making it a vital component in the development of infrastructure (Cuenca & Ferrara, 2017). When it comes to Fiber-Reinforced Cementitious Composites (FRCC), a great deal of focus is placed on factors like the type of fibers used, the amount of fibers in the mixture, variations in the matrix, and the methods used for processing. The characteristics of the composite are shaped by these elements, resulting in enhanced mechanical durability and bond properties that are essential for maintaining structural integrity (Li, 2003).

High-strength cementitious Composites are a revolutionary advancement in construction materials, surpassing the limitations of traditional concrete. These composites are designed to meet the needs of modern construction tasks, where durability and long-term performance are crucial. It includes everything from strong concrete to innovative DSP materials. ECC is a remarkable material that demonstrates resilience and adaptability when confronted with ever-changing environmental conditions. Thanks to its remarkable ductility along with crack control abilities, ECC provides a revolutionary solution for infrastructure uses that demand unmatched durability and performance (Pacheco et al., 2018).

In addition, this chapter explores the field of sustainable construction using High-Volume Fly Ash Concrete. It focuses on utilizing industrial by-products to minimize environmental harm and improve material characteristics. In the meantime, Structural Lightweight and Heavyweight Concrete provide customized solutions for a wide range of construction needs, such as decreasing dead load or meeting specific project demands (Siad et al., 2018).

6.1. FIBER-REINFORCED CEMENTITIOUS COMPOSITES

Learning Objectives

- Understand the role of fiber volume fraction in influencing mechanical properties of fiber-reinforced cementitious composites.
- Identify the different fiber types commonly used in FRC and their respective characteristics.
- Recognize the effect of processing methods on the mechanical properties and density of fiber-reinforced cementitious composites.

Fiber-reinforced cemented composites are cement materials with fibers mixed in. The fibers are usually short and broken. Fiber-reinforced concrete (FRC) has been around for a long time, but it started to be used more and more in the 1960s. FRC was mostly created to fix two major problems with cement-based composites: containing no very high tensile strength and not very tough or energy-efficient. The roles of fibers can be divided into cement-based compounds into two groups: Controlling cracks caused by shrinking and improving mechanical properties. It is common practice to control shrinkage cracks to incorporate small quantities of fibers with low modulus and strength. These fibers serve to limit early-age shrinking and minimize the occurrence of shrinkage cracking. To enhance mechanical properties, reinforced fibers have been widely adopted in different concrete structures. This technique aims to enhance flexural performance, boost impact resistance, and potentially alter the failure mode (Joo Kim et al., 2008). The addition of fiber greatly impacts the mechanical characteristics and failure mechanism of FRC. There are three groups in which fiber-reinforced cementitious composites might be classified, depending on the amount of fiber added. Low-volume fiber fractions (<1%) are commonly used in FRCs to help minimize shrinkage cracking. Fiber-reinforced composites with mild volumetric fiber fractional ones (between 1 and 2%) demonstrate enhanced mechanical characteristics, such as increased Modulus of Rupture (MOR), improved fracture toughness, and enhanced impact resistance. The fibers within this type of FRC can serve as secondary reinforcement within structural components, such as partially replacing shear steel stirrups or controlling crack width in constructions. Over the past few decades, a new type of FRC, known as high-performance FRC or HPFRC, has emerged (Savastano et al., 2009). High-performance fiber-reinforced composites demonstrate a remarkable ability to withstand strain and strengthen due to their utilization of a high concentration of fibers. Several types of high-performance fiber-reinforced composites are used in cementitious materials. These consist of SIFCON, which involves the use of steel fibers in the concrete mix; SIMCON, a variation of SIFCON that uses a steel fiber mat rather than discrete fibers; and CRC, which utilizes fine, strong, and stiff steel fibers along with densely organized reinforcing bars to create a strain-hardening matrix. HPFRC typically has a tensile strain ability of around 1.5% (Abdallah et al., 2018).

In typical scenarios involving fiber-reinforced cementitious composites, where the amount of fibers used is relatively small (e.g., 0.5% fibers from steel or 0.05% polypropylene fiber), the role of the fibers becomes noticeable only when a significant crack develops in the composite. Despite just one major crack and the composites exhibiting strain softer after reaching the peak load, the inclusion of fibers results in a notable enhancement in the overall energy absorption and toughness of the composites. This is demonstrated by the increased area under the strain and stress or load-displacement curve, which is depicted in Figure 6.1 (Song et al., 2021). In certain situations, assuming there have been no fiber fractures, the process of fiber debonding along with pullout can result in significant energy consumption. However, when the volume of fibers is increased, the micro-cracks that develop in the matrix can be stabilized. This is because the matrix and fibers interact and bond together, which delays the development of the initial significant crack within the matrix. Therefore, the perceived elasticity of the matrix may be enhanced.

In addition, the inclusion of an important fraction of small-diameter glass, stainless steel, or synthetic fibers within a cement-based matrix may result in strain-hardening along with multiple cracking actions. This interaction between the fibers and matrix alters the failure mode from quasi-brittle to pseudo-ductile. Consequently, the toughness of the composites can be greatly enhanced, along with improvements in deformation capability and tensile strength. One possible mechanism for reducing the propagation of a transverse crack within unidirectional fiber composites involves the formation of the longitudinal cylindrical shear small cracks at the interface among the fibers along with the matrix. These micro-cracks enable the fibers to separate from the matrix while still transmitting the load across the surfaces of the main crack. Fibers can enhance the durability and elasticity of cement-based composites, while also improving their bending resistance. Nevertheless, the inclusion of fibers has a limited impact on the compressive capacity of composites made up of cement. When the fiber volume fraction is small, the impact is minimal. Therefore, it is not beneficial to include fibers to improve compressive strength. The hierarchy of fiber influence on the mechanical characteristics of composites made from cement is as follows: toughness is the most significant, followed by flexural strength, tensile strength, and finally compressive strength. While fibers can enhance the mechanical characteristics of materials made from cement, it is crucial to acknowledge that, in most cases, fiber reinforcement cannot replace traditional reinforcement or rebars. Fibers along with rebars have distinct functions in modern concrete technology and are commonly employed together in various applications, including CRC and certain crucial structural elements or connections (Yoo et al., 2013).

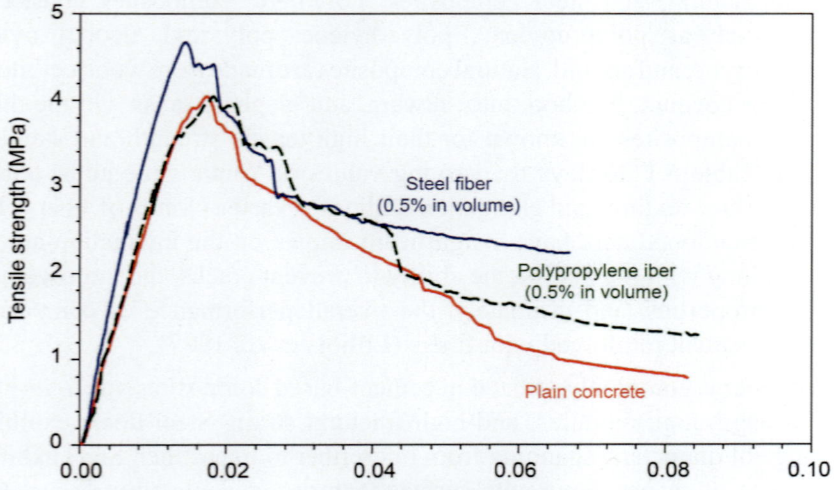

Figure 6.1. *Illustration of load-displacement curves of plain and fiber-reinforced concretes.*

Source: Ainger S. Creative Commons License.

6.1.1. Determinants of the Characteristics

Several factors can affect the characteristics of fiber-reinforced cement-based materials made of composites including the type and amount of fibers, variations in the matrix, and the processes used in manufacturing. This section provides a detailed discussion of these parameters (Kanakubo, 2006).

6.1.1.1. Fiber Type

There are various ways to evaluate the fiber type. When considering size, fibers can be divided into two categories: macro- and microfibers. The diameter of larger fibers typically falls between 0.2 to 1 mm, while smaller fibers range from a couple of to tens of micrometers within diameter. In the realm of academic writing on cementitious composites, it is worth noting that microfibers have proven to be highly effective in preventing the formation of microcracks, while macro fibers have shown great potential in preventing the occurrence of larger, macroscopic cracks (He et al., 2001). Regarding the materials employed for fiber-reinforced cement-based composites, commonly utilized fibers include carbon and various types of glass (such as borosilicate and alkali resistant), There are various types of materials used in the construction industry, including polymeric, natural, and steel composites. Polymeric composites consist of materials such as polypropylene, polyethylene, polyvinyl alcohol nylon, polyester, acrylic, and aramid. Natural composites are made from wood cellulose, sisal, coir or coconut, bamboo, jute, akwara, and elephant grass. On the other hand, steel composites are known for their high tensile strength and stainless properties. Table 6.1 displays the varying values of Young's modulus, tensile strength, surface texture, and elongation ability for various kinds of fibers. The properties mentioned here have a significant impact on the interaction among the fibers along with the matrix, the ability to prevent cracks, the improvement of matrix properties, and ultimately, the overall performance of composites made from cement reinforced with fibers (Lillioja et al., 1987).

Steel fiber is commonly utilized in cement-based composites due to its high tensile strength high modulus, and high fractural strain. Steel fibers exhibit a wide range of diameters, spanning from microfiber to microfiber. Steel exhibits a wide range of shapes along with surface textures, as depicted in Figure 6.2. Steel fibers with textured surfaces and unique shapes enhance the bond strength. An issue with steel fibers is their high specific gravity, leading to an increased dead load within the composite.

Fiberglass is a frequently utilized material in FRC, typically in the shape of filaments. There are a total of 204 filaments in each bundle of glass strands (Korhonen et al., 2009). Glass fibers are typically provided in the form of ongoing roving and may be easily chopped into shorter fibers using a readily available chopper. Glass fibers exhibit a remarkable tensile strength of 2.1 – 2.5 GPa and possess a high fracture strain. However, their modulus is relatively low, measuring around 70 GPa. In addition, it is important to note that regular borosilicate glass fibers (E-glass) and soda-lime glass fibers (A-glass) are susceptible to damage from alkali solutions in cement-based composites. As a result, their durability is compromised, and it is advisable to exercise caution when using them. AR glass fibers are formulated with approximately 16 – 20% zirconia (ZrO_2) to protect against high-alkalinity attack. AR fiberglass is widely utilized in composites made from cement due to its popularity. Glass fibers have certain drawbacks, such as their limited ability to withstand cyclic loads moisture, and sustained loads (Verdijk et al., 2010).

Table 6.1. *Properties of Different Types of Fibers*

Fiber	**Diameter (µm)**	**Specific Gravity**	**Tensile Strength (GPa)**	**Elastic Modulus (GPa)**	**Fracture Strain (%)**
Steel	5 – 500	7.84	0.5 – 2.0	210	0.5 – 3.5
Glass	9 – 15	2.6	2.0 – 4.0	70 – 80	2.0 – 3.5
Fibrillated polypropylene	20 – 200	0.9	0.5 – 0.75	5 – 77	8.0
Cellulose	—	1.2	0.3 – 0.5	10	—
Carbon (high strength)	9	1.9	2.6	230	1.0
Cement matrix for comparison	—	2.5	3.7 × 10−3	10 – 45	0.02

Source: Ainger S. Creative Commons License.

Carbon fibers are known for their superior strength and stiffness compared to glass fibers, although come at a higher cost. Carbon fibers exhibit a wide range of strength, modulus, and ultimately strain values. The strength can vary from about 2.2 to 5 GPa, the modulus about 800 to 250 GPa, along with the strain at its highest point of 0.3 to 1.8% (Hikida et al., 2000). The specific gravity varies depending on the modulus of the fiber, with a range of 2.2 for high-modulus fibers to 1.8 for low-modulus fibers. There is a correlation between the stronger fibers and their lower moduli along with higher final strain. In addition to their impressive stiffness and strength, carbon fibers also demonstrate exceptional durability against moisture or chemicals and are not affected by fatigue. Carbon fibers have certain limitations, such as their susceptibility to impact, limited ability to withstand strain, and relatively high cost (Burkholder et al., 1994).

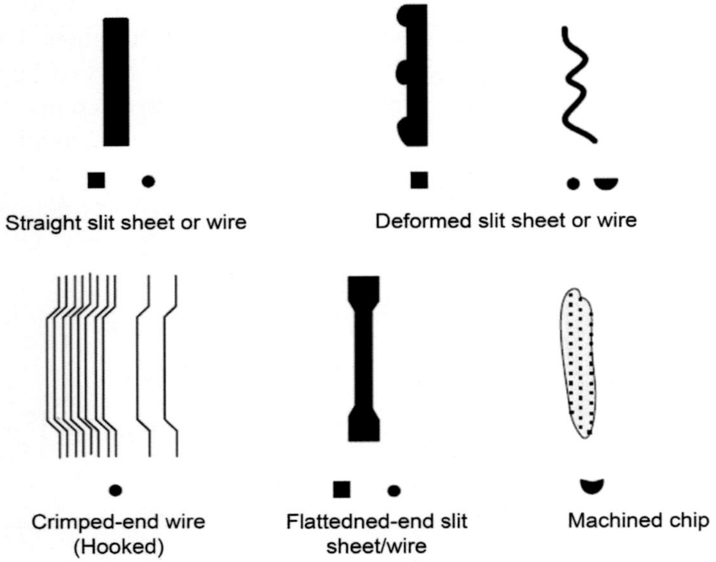

Figure 6.2. *Illustration of different shapes and surface textures of steel fibers.*

Source: S Charles, Creative Commons License.

6.1.1.2. Volume Fraction of Fiber:

The fiber volume percentage ratio is a crucial factor that has a significant impact on the properties of FRC. It is determined by calculating the ratio for each fiber volume relative to the entire volume of FRC. As previously stated, when the fiber volume ratio is low, the inclusion of fibers primarily affects the energy-consuming characteristic. When the percentage of fiber volume ratio is increased, it is possible to improve the shear strength within the matrix and alter the mode of failure (Fu et al., 2000).

In cases where the fiber volume fraction has become low, the final stress of Fiber Reinforced Cementitious Composites (FRC) is typically attained when the matrix itself attains its ultimate stress level. Based on the rule about mixtures, the final strain of the composites will be expressed as

$$\sigma_{cu} = V_m \ \sigma_{mu} + V_f \sigma_f$$

$$= V_m E_m \varepsilon_{mu} + V_f E_f \varepsilon_{mu}$$

$$= V_m E_m \varepsilon_{mu} + V_f \frac{E_f}{E_m} \sigma_{mu}$$

Where,

- σ_{cu} = ultimate stress in fiber-reinforced concrete
- σ_{mu} = ultimate stress in the matrix
- σ_f = stress in the fiber
- E_m = Young's modulus of the matrix
- ε_{mu} = ultimate strain
- V_m = matrix volume fraction ratio
- V_f = fiber volume fraction ratio (Dhakal et al., 2007).

6.1.1.3. Matrix Variation

The characteristics of the matrix play a crucial role in determining the bond strength with the fibers along with the mechanical characteristics of Fiber Reinforced Composites (FRC), including the ultimate tensile strength.

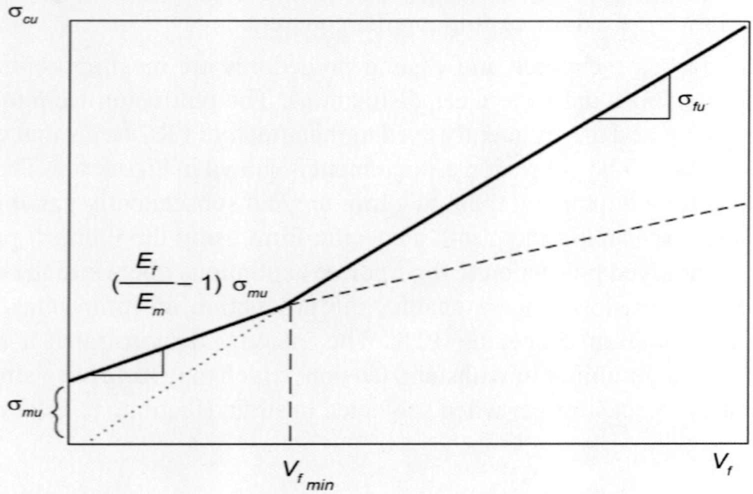

Figure 6.3. *Description of relationship between stress and fiber volume fraction.*

Source: Roberto L. Creative Commons License.

Mineral admixtures, including silica fume, fly ash, slag, and metakaolin, can be used to modify the FRC matrix. Additionally, the composition can be altered through the incorporation of water-soluble polymers such as PVA. Modifying the composition of the matrix may improve the bond characteristics with the fibers, bolster the toughness of the matrix, and strengthen the matrix's tensile strength, thereby improving the overall mechanical properties of Fiber Reinforced Composites (FRC) (Zhou et al., 2021).

6.1.1.4. Processing Methods

Recent research has indicated that the behavior of composites reinforced with fibers is influenced by the processing techniques employed. Various industrial processing methods are commonly employed to produce FRC products. These methods consist of the extrusion process known as the Reticem process, the pultrusion process, the Hatschek process, and normal casting. Various processing techniques result in FRC products that exhibit varying densities, bonds, and consequently, distinct mechanical characteristics.

The Hatschek technique was first designed for the production of FRC products containing asbestos fibers through a dewatering procedure, as depicted in Figure 6.4 (Mir et al., 2017). Nevertheless, with the increasing number of countries imposing limitations on the utilization of asbestos fiber for health reasons, the technique is primarily employed for the production of thin-wall items, like sidings, utilizing cellulose fibers. The conventional casting procedure is identical to that used for casting regular concrete.

The Hatschek technique and casting procedures are most appropriate for fibers that are short and have clear distinctions. The pultrusion technique and the Reticem method are frequently used to manufacture FRC items that contain continuous fibers. The pultrusion arrangement is shown in Figure 6.5. The fibers are reinforced by immersing them in a mixture and subsequently passing them through a die or shaping them into a specific form using the filament process. The study employed pultrusion to incorporate continuous fibers into the cement matrix. The pultrusion process enables the production of composites with a fiber volume fraction exceeding 10%. The material demonstrated a notable augmentation in its ability to withstand tension, which may have been attributed to its capacity to get stronger when subjected to strain (Iaquinto & Fredrickson, 1997).

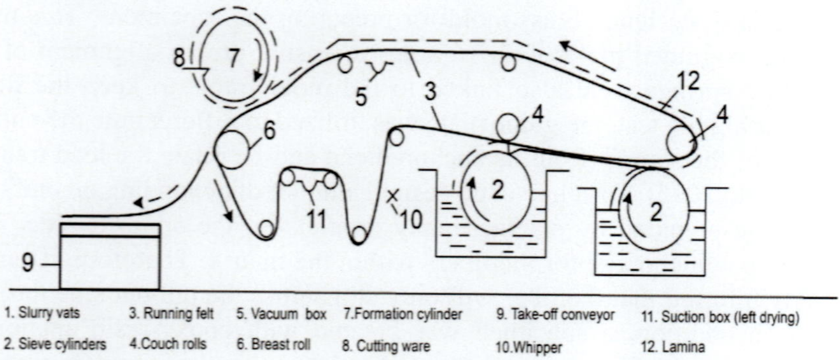

Figure 6.4. *Illustration of the hatschek process.*

Source: L. Mote Creative Commons License.

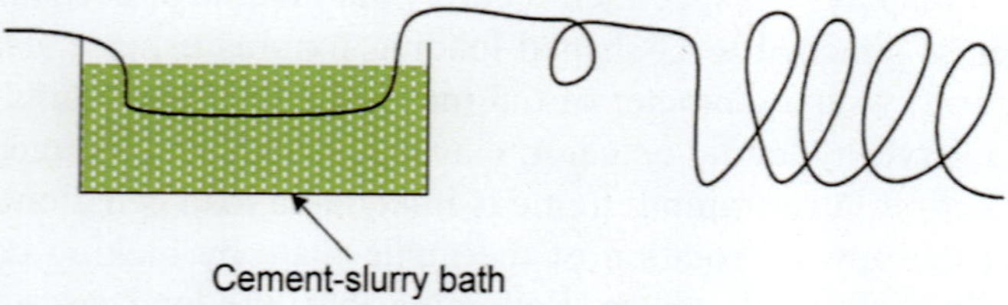

Figure 6.5. *Illustration of the pultrusion process.*

Source: Carlos JR. Creative Commons License.

6.1.2. Fiber-Cement Bond Properties

Extensive research has been carried out to analyze the properties of the bond among the fiber along with the cement matrix in FRC, as well as the behaviors of fiber debonding and pullout, due to their crucial role. In the realm of material characteristics for fiber debonding and pullout issues, two methods of analysis have emerged: energy release rate-based approaches and stress-based ones. Based on the stress-based criterion, the fiber and matrix separate when the highest value of shear stress during the interface attains a critical value. Based on the energy amount released criterion, debonding occurs when the energy entering the interface surpasses a particular resistance energy threshold (Beglarigale et al., 2016).

Measuring the bonding characteristics of the fiber-cement matrix can be done through pullout testing. Creating the specimens for this test may represent a complex process, with various methods available. In their work, Rotter B.

used a specially designed brass mold for preparing the specimens. This mold included brass-guided plates with 16 holes to ensure proper alignment of the fibers. Small springs were also linked to the mold frame to keep the fibers directly. During the test, the guide plate was utilized to differentiate the pulled-out section of the sample from its anchored end and facilitate the load transfer (Ferreira et al., 2015). To achieve the desired amount of debonding on one side, samples were prepared by placing epoxy droplets on the opposite side. This ensured proper anchorage for the fibers within the matrix. Therefore, it can be reasonably inferred that the fiber will only slip across the pullout side through the tests. In addition, a specimen was created with epoxy-resin anchorage offered on each side. The specimen was utilized to assess the fiber deformations within the guide plate. This measurement was then subtracted from the total slip displaced of the pullout samples, as detailed in the subsequent sections (Shah & Li, 2003).

Figure 6.6 displays the specimen securely held within a mechanical fixture. The specimen is attached to U-shaped loading fixtures using a stainless-steel rod (loading rod) with a diameter of 6.3 mm (0.25 in.). The loading fixture is attached to a servo-hydraulic actuator, causing the whole specimen fixture to move along with it. A restraining frame is linked to a load cell along is utilized to counteract the upward rotation of the guide plate by making contact with the specimen's brass guide plates. Following that, the load passes through a brass guide plate into the matrix. The fibers play a crucial role in linking the upper and bottom portions of any specimen, effectively reacting contrary to the applied load. The load gets moved to the highest point of the sample and then to its loading fixture by bending the loading rod (Lima et al., 2017).

Two extensometers were utilized to measure the slip displacement caused by fiber debonding and pullout. These extensometers ranged from 1.905 mm (or 0.075 in.) and were mounted throughout the guide plate. They had a gauge length of 12.5 mm (or 0.5 in.). The feedback signal for controlling the servo-hydraulic mechanism was obtained by using the mean output of the extenders. The pullout tests were conducted with a gradual revealing of the extensometer transducers at a rate of 0.0254 mm per minute. After surpassing the transducer's range, the control mode shifted to stroke control, resulting in the complete extraction of the fiber length from the matrix (Bezerra et al., 2004).

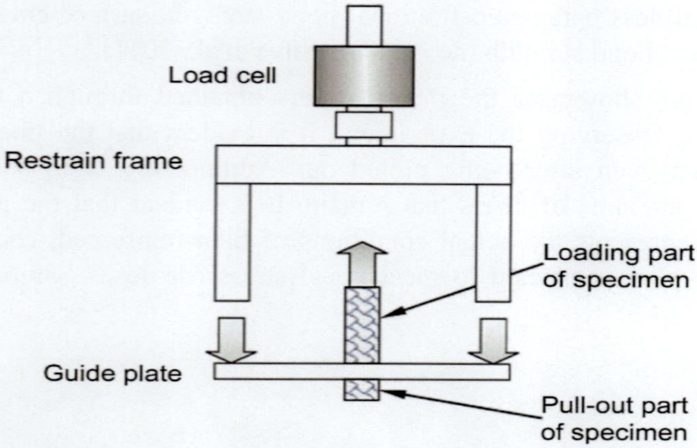

Figure 6.6. *Illustration of specimen being held in the mechanical fixture.*

Source: Bhaskar Kumar, Creative Commons License.

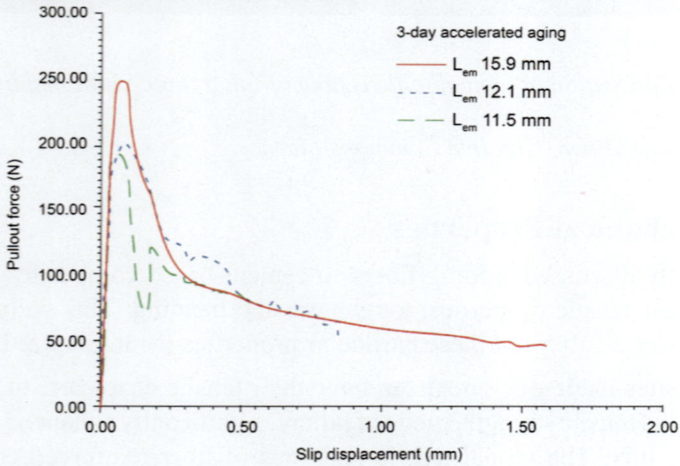

Figure 6.7. *Experimental results of fiber pullout test.*

Source: Arther Lynos, Creative Commons License.

Figure 6.7 illustrates a standard plot of pullout load slip displacement. Figure 6.8 clearly illustrates the initial linear relationship between the load and displacement. Afterwards, there is a transition to unpredictable behavior till the load limit is reached. Subsequently, there is an abrupt decrease in the load, by a gradual reduction in the load as displacement occurs. This indicates the occurrence of fiber pullout via the matrix. Through the analysis of the key points on the curve, it is possible to interpret the characteristics of the interface, including stiffness parameter, frictional bond strength, surface energy release rate, and shear bond strength (de Andrade Silva et al., 2011).

Figure 6.8 showcases the pullout fibers obtained through a three-point bending test. Observing the experiment, it is evident that the fibers of steel remain intact even after being pulled out. Additionally, both sides exhibit comparable amounts of fibers that remain. It is evident that the pullout test accurately represents the actual conditions of fiber-reinforced, cement-based composites when subjected to mechanical stress (de Jesus Santos & Lima, 2015).

Figure 6.8. *Illustration of typical pullout fibers from a three-point bending test.*

Source: Barnali Dixon, Creative Commons License.

6.1.3. Mechanical Properties

As previously discussed, adding fibers to cement-based composites primarily enhances their tensile properties, toughness, and bending. This section delves into the impact of fibers on these particular properties (Mobasher & Li, 1996).

Composites made of cement can have their tensile characteristics, such as toughness, and tensile strength, mode of failure, significantly enhanced by adding fibers to the mix. The global tensile reactions of fiber-reinforced composites made from cement are typically characterized by either strain softening or strain hardness. Commonly used to enhance strain-softening fiber-reinforced cement-based composites are small volumes of relatively short fibers. In bulk usage involving large quantities of concrete, these types of composites which contains around 1% fiber are commonly utilized. With the addition of a low-volume fiber percentage, the tensile reactions of these composites are shown in Figure 6.1. Stress deform curves for a simple concrete specimen are also displayed for comparison (Huang et al., 2021). The dimensions of the fibers of steel for example of concrete reinforced with steel fiber were 48 mm in length, 0.5 mm in diameter, and 0.5% by volume fraction. Fibrillated polypropylene fibers used to strengthen concrete specimens should be 50 mm in length and

0.5% by volume. Figure 6.1 clearly illustrates that once the peak load has been attained, all the specimens exhibit a softening behavior. However, it is worth noting that the fiber-reinforced specimens have larger areas according to the load's deformation curve compared to the plain concrete specimens. It can be inferred that the specimens reinforced with fibers demand a greater amount of energy for fracture compared to the standard concrete specimens. In the case of strain-softening failure, typically only a single prominent crack develops in the specimen. The material's fracture procedure can be categorized into four stages. First is the linear elastic stage, which occurs at the beginning and accounts for 0 – 35% of the peak load. Next, the dispersed damage stage is attained, which takes place between 35% and approximately 80% of the peak load within the pre-peak stage. Following that is the micro crack localization stage, which occurs throughout the loading period among 80% of the pre-peak load and 80% of the post-peak load. Finally, the major crack propagation stage is attained, which happens after 80% of the post-peak load is reached (Yang et al., 2009).

When using a high-volume portion of short fibers or aligned continuous fibers, a failure mode involving several matrix cracks can occur in cement-based composites. These types of composites have found applications in various forms such as thin sheets for curtain walls, slurry-filled composites, along with concrete. Figure 6.9 displays a strain-hardening curve commonly observed in steel fiber-reinforced cementitious composites. At a specific point on the curve, denoted as B, there exists a bend-over point (BOP) where the matrix's impact on the tension capability reaches its peak. The stress-strain curve depicted in Figure 6.9 can be approximately categorized into four distinct stages (Kim et al., 2017). Stage 1 in the figure represents the initial phase, where the composite exhibits elastic behavior. Towards the end of this stage, some micro cracks start to form and are scattered randomly throughout the material. The second stage spans from point A to point B. During this phase, micro-cracks that were initially scattered throughout the material begin to concentrate and develop into a significant crack that spans the entire cross-section of the specimen at point B. The BOP signifies the culmination of phase 2. The third stage takes place among points B and C. At this stage, fibers take on the additional load at the site of the first crack, which is then transferred back to the matrix through the bond. This repetitive process results in the restoration of tensile stresses inside the matrix (Sahmaran et al., 2010).

When the tensile stress exceeds the strength of the matrix, an additional crack will form in the matrix. The cracking process continues until the cracks reach their minimum spacing. At this point, the tensile stress is transferred from the fibers back to the matrices and is unable to exceed the breaking strength of the matrix. The process leads to a uniform distribution of tiny cracks all over the matrix. Point D is reached during the fourth stage. It is unlikely that any additional cracks will develop in the matrix at this stage. The fibers will support the extra load until they ultimately give way. One can observe the presence of strain hardening and multiple cracks in various kinds of concrete, such as those that are reinforced via continuous fibers with short fibers (Parveen et al., 2013).

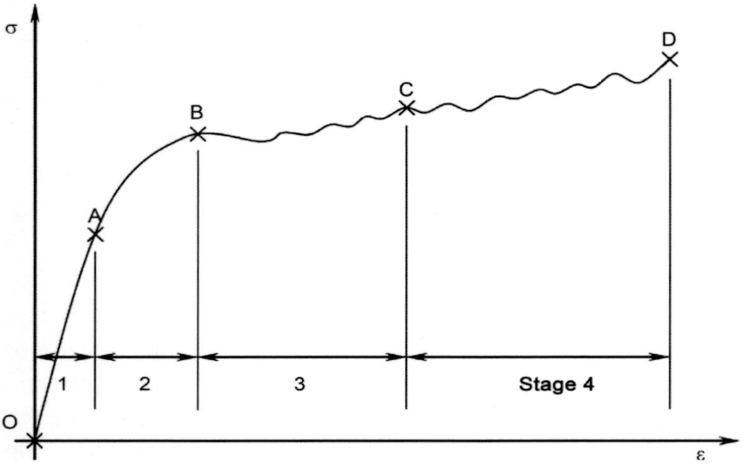

Figure 6.9. *Illustration of Strain hardening behavior of steel fiber-reinforced cementitious composite.*

Source: Karen Huss, Creative Commons License.

6.2. CEMENTITIOUS COMPOSITES OF HIGH-STRENGTH

Learning Objectives

- Learn the key factors and methods for achieving high-strength concrete.
- Understand the properties and advantages of MS and DSP materials.
- Explore the applications and considerations of DSP materials in construction.

6.2.1. Concrete of High-Strength

Concrete with a compressible cylindrical strength greater than 50 MPa is considered high-strength concrete, which is made using normal-weight aggregates. In this context, high strength refers specifically to high compressive strength. While the raw materials for normal and high-strength concrete are the same, producing the latter requires stricter quality control and more meticulous material selection (Zhang et al., 2021).

Typically, when selecting aggregate for high-strength concrete, it can be preferable to opt for one that possesses an elevated crushing strength, if available. Typically, the permitted size of aggregate can be limited to 20 mm. The restriction on the maximum aggregate size aims to minimize the impact of the area of transition and achieve a more uniform material. Accurate calculation of moisture levels in aggregates is crucial to ensure the attainment of the appropriate water-to-cement or water-to-binder ratio. The material percentages of high-strength concrete vary from those of normal-strength concrete construction. The cement content typically falls within a range of 400–600 kg/m^3. The higher amount of cement is due to the restriction on the largest possible aggregate size along with the requirement for workability according to the smaller water-to-cement ratio condition, as we will explore further. In addition, increasing the amount of cement results in a more uniform composition of the concrete. It is common to include water-reducing along with mineral admixtures silica fume, and fly ash, like slag, in high-strength concrete mixtures (Ding et al., 2022).

When making high-strength concrete, the ratio of water to cement or water to binder is essential. It is evident from Abrams' law and the previously discussed gel/space ratio that less of a water-to-cement and water-to-binder ratio can lead to higher compressive strength. Therefore, a common approach in producing high-strength concrete was to decrease the water-to-cement or water-to-binder ratio from the typical values used for regular-strength concrete (approximately 0.5 or higher) to around 0.3 or less. Nevertheless, reducing the water-to-cement or water-to-binder ratio can result in a decrease in workability, particularly fluidity. An effective solution to this drawback in concrete proportions is to incorporate super plasticizers and enhance the cement content. As previously mentioned, a super plasticizer can separate cement particles, allowing for the release of entrapped water. This separation can be achieved through either electrostatic attraction or steric hindrance. Therefore, it is possible to enhance the workability of concrete. By expanding the cement content, the total quantity of water utilized by the concrete mix can be increased, resulting in more paste to produce the lubrication effect. This, in turn, enhances the workability of the mixture (Prabha et al., 2020).

Including mineral admixtures, particularly silica fume, is a crucial element in producing concrete with exceptional strength. It is worth noting that silica fume seems a byproduct generated by the ferro-silicon production method and is characterized by its fine particle size and strong reactivity. Because of its tiny size of approximately a hundred nanometers, silica fume can tightly fill the spaces between cement particles, which are typically around 20 microns in size. This results in the formation of a significantly more compact microstructure. Because of its highly reactive and amorphous nature, the fume of silica reacts using calcium hydroxide (CH, formed during cement hydration) and water (Khan, 2012). This reaction results in the formation of secondary C–S–H, which effectively fills large capillary spaces. As a result, the density within the microstructure is enhanced and the porosity of the concrete is reduced. As a result, the concrete's strength and permeability are improved. In addition, the decrease in CH and permeability results in a reduction in the rate and quantity of moisture, chloride, carbon dioxide, and oxygen that can penetrate the concrete. High-strength concrete is less susceptible to corrosion, sulfate attack, leaching, and alkali-aggregate reaction carbonation (AAR), among other variables that can shorten its lifespan. The overall durability of the concrete is so improved. The responses of these two concretes to compressive load are shown in Figure 6.10. Compared to normal-strength concrete, high-strength concrete develops tension at a substantially faster pace. This implies that high-strength concrete demonstrates high strength from an early age (Sohel et al., 2018).

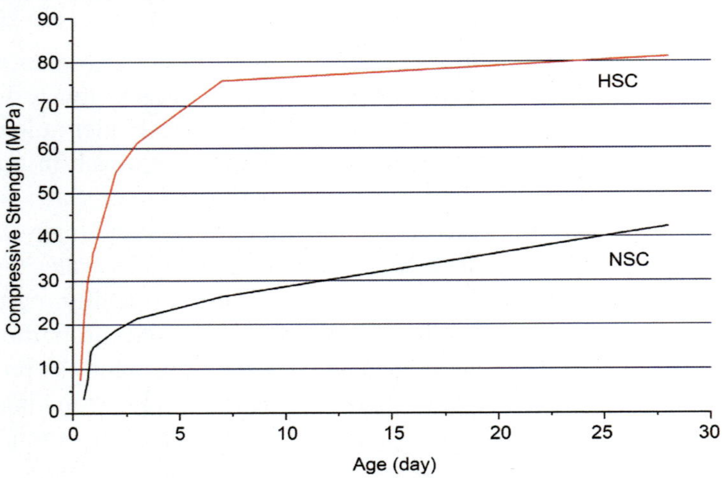

Figure 6.10. *Illustration of Compressive stress development of normal-strength concrete (NSC) and high-strength concrete (HSC).*

Source: Tarun R. Creative Commons License.

6.2.2. MS Concrete

Microsilica, commonly referred to as MS, is a crucial component in MS concrete, typically comprising up to 15% of the total weight of cement-based substances. There are several MS concretes that can be found in the US as well as Europe. The microstructure and attributes of MS concrete exhibit distinct characteristics (Ranade et al., 2013).

The presence of silica within MS concretes leads to a significant reduction in the amount of CH, thanks to the pozzolanic reaction. The formation of C–S–H in MS concrete, when reactive pozzolanic material like amorphous micro silica is present, exhibits slight variations compared to the C– S– H formed during the regular hydration process of cement that was (Liao et al., 2019). Specifically, the C/S proportion is decreased from approximately 1.7 to around 1.1, while the silicate structure remains largely unchanged. Both types, "normal" and "pozzolanic," are likely to co-exist, resulting in an average value of approximately 1.5. By incorporating microscopic silicon dioxide, the permeability can be significantly reduced or even eliminated by targeting the macro porosity with a diameter larger than 100 nm. Therefore, MS concrete has the potential to be both impermeable and long-lasting. By incorporating significant quantities of microscopic silicon dioxide, the interface zone of MS concrete acquires a more dense and uniform structure, resembling that of the bulk paste. According to Ma and Li (2014), the fine particles of micro silica can prevent the formation of water film around the aggregate particles and reduce the negative impact of binder-packing particles on the aggregate surface. These factors are believed to contribute to the formation of porous interfacial zones of transition. In a similar vein, the connection between the paste and other materials that are embedded and have hydrophobic surfaces, like glass and steel fibers, will also experience enhancement (Sattarifard et al., 2022).

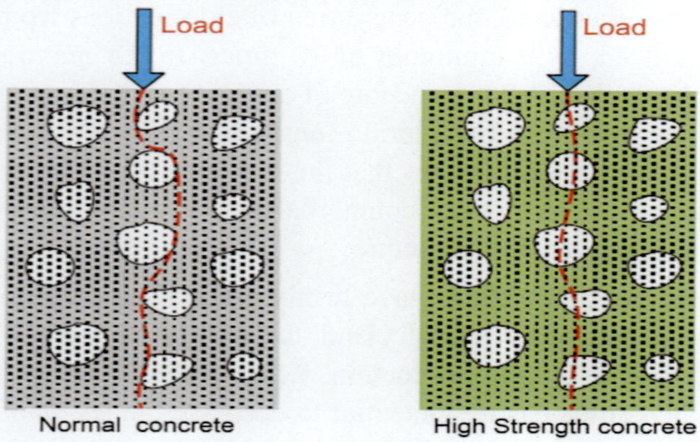

Figure 6.11. *Illustration of the failure modes of normal-strength concrete (NSC) and high-strength concrete (HSC) under compression.*

Source: Ainger S. Creative Commons License.

It has been well-established for quite some time that micro silica is highly effective in decreasing the permeability of concrete. Research has demonstrated that by incorporating micro silica, modest permeability coefficients can still be achieved regardless of low amounts of cement and high w/c ratios. The efficiency aspect is approximately 10 when incorporating 10% silica, meaning that replacing 1 part cement with 1 part small particles of silica was similar to introducing 10 parts cement. Put simply, adding 10% micro silica can reduce permeability to a level similar to increasing the cement content. This evaluation is centered on properly cured concretes. Insufficient curing can significantly harm MS concrete compared to conventional concrete, as it can hinder the pozzolanic reaction if not properly cured (Choi et al., 2022).

The enhanced pore structure enhances the chemical durability of MS concrete and limits the infiltration of oxygen and other harmful substances. The reduction in alkalinity resulting from the reaction of pozzolanic acid is not substantial. A study conducted by Diamond (1985) observed that the pH level exceeded 12 after 145 days of hydration within a concrete mixture containing 30% micro silica. For more than two decades, MS concrete in Kristiansand, Norway, has withstood the harsh effects of salt spray without showing any visible signs of corrosion (Lei et al., 2021).

When chlorides are introduced into MS concrete, the binding of chloride ions through the hydration products seems to be less pronounced compared to conventional concretes. This is because there is a decrease in the amount of aluminates available to react using chloride ions and form Friedel's salt. Therefore, the de-passivation of steel may occur with lower amounts of added chlorides. On the other hand, the penetration of chloride ions from the outside is greatly minimized, by a minimum of 10 times. In the realm of academic writing, it has been observed that the rate of corrosion remains consistent when exposed to equal quantities of chloride ions, irrespective of the micro silica content in the concrete. This means that the presence of micro silica does not have a detrimental effect that can counterbalance or overshadow the positive impacts of a more refined pore structure.

Multiple independent studies have firmly established the effectiveness of micro silica in managing AAR. In Iceland, the addition of 7.5% Microsilica is a common practice in cement production. Research has demonstrated that the inclusion of Microsilica has the potential to decrease the levels of alkalis ($K+$ or $OH-$) within the pore solution. The enhanced impermeability and decreased CH content are expected to enhance the durability against sulfate along with acid attacks. Research has shown that micro silica outperforms other pozzolans when added in equivalent amounts. Extensive field trials conducted in Norway's capital city of Oslo have demonstrated excellent long-term performance under challenging conditions (Tran & Kim, 2014).

6.2.3. DSP Materials

DSP is an acronym that means densified via small particles. DSP concrete shares a close resemblance in makeup to MS concrete. It is logical to anticipate that the performance of DSP contents can be predicted based on the properties of MS concrete. While there may not be a distinct distinction between MS concrete as well as DSP substances, DSP materials possess certain distinct characteristics: (1) DSP concretes typically have a significantly lower water-to-binder ratio, around 0.20. (2) DSP concretes contain a much higher percentage of micro silica, exceeding 20% by weight. (3) The particle packing requirements for DSP concrete are more demanding (Pan et al., 1998).

It is crucial to achieve an optimal packing of particles for DSP concrete, so the selection of aggregate grading and binder particle size distribution must be done with great care. The aggregates typically employed for DSP concrete tend to be unique. For example, complex stainless steel, granite, diabase, along calcine bauxite aggregate are commonly utilized.

Given the highly dense arrangement of particles and the low water-to-binder ratio, a significant amount of dispersing agent, such as a super plasticizer, must be employed. At such low water-to-binder ratios, approximately half of the cement can undergo hydration. Thus, it can be estimated that DSP concrete consists of around 50% unhydrated cement particles. The particles serve as a strong base to uphold the hydrates and withstand the compressive load (Prakash & Kumar, 2023). Significant amounts of silica fume greatly reduce the creation of CH within forms of hydration products, thanks to its pozzolanic reactivity. The presence of large, solid crystals of CH is not observable, and its microstructure is consistently uniform compared to regular cement paste. The degree of porosity of DSP concrete has a significantly low level, and the structure of its pores exhibits notable distinctions compared to conventional Portland cement concrete, just like illustrated in Figure 6.12. DSP materials do not contain any capillary pores greater than 100 nm (Ravitheja et al., 2021).

The strength of DSP concrete develops quickly, with an initial setting time ranging from 15 – 20 hours. Within approximately 48 hours, the compressive property of DSP concrete may exceed an impressive 100 MPa. Extended moist curing can lead to further increases in strength. The highest stated measurement for DSP materials is a compressive force of 345 MPa. This outcome is achieved by eliminating micro porosity and creating a densified microstructure. Ensuring a proper distribution of the tiny particles of silica is crucial. When low-shear mixing is employed, relying solely on a chemical-based dispersion agent is not enough. By utilizing high-shear mixing, a more effective distribution of Microsilica can be achieved, resulting in a notable enhancement in compressive strength. Through careful optimization of the processing techniques, it is feasible to attain paste and mortar abilities exceeding 200 MPa (Guerrini, 2000).

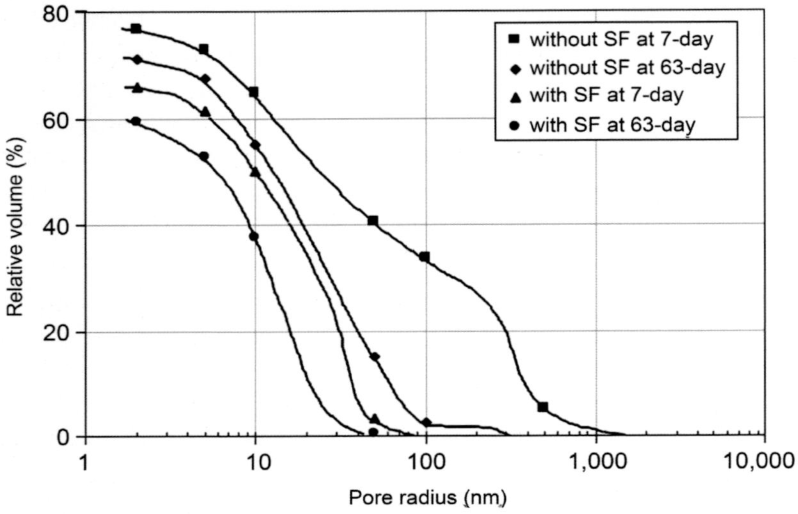

Figure 6.12. *Description of the effects of the addition of silica fume on pore size distribution of DSP materials.*

Source: Roberto A Creative Commons License.

Similar to high-strength concrete, DSP content exhibits a high level of brittleness. The compressible-to-flexural strength ratio for paste typically falls within a range of 10 – 12:1, while for mortar it is generally between 7 – 12:1. Therefore, it is necessary to incorporate reinforcement to enhance flexibility. In terms of extra physical attributes, the abrasion resistance of DSP material is remarkable. A screw feeder manufactured from DSP material for fly ash has a stated lifetime of 1250 hours, which is much longer compared to a feeder made from steel with a wear-resistant carbide layer, which has a lifespan of only 250 hours. It is anticipated that the permeability of DSP content would be exceptionally low due to the optimized particle packing, and low water-to-binder ratio, along with the pozzolanic reaction that occurs in products containing high-dose micro silica. The commercial goods DASH 47 has demonstrated an air permeability that is substantially lower than the equivalent amount of cast aluminum (Jamal Shannag & Hansen, 2000).

Typically, DSP materials find application in the creation of molds and tools. The durability of DSP is remarkable, as it can maintain its properties even when exposed to temperature cycling of up to 200 – 250°C. When the temperature exceeds a certain limit, significant deterioration in the mechanical characteristics occurs. To surpass this limitation, the technique of impregnation via high-temperature polymer compounds has been employed. Experimental evidence demonstrates that the flexural strength can be greatly enhanced and the leaking rate can be reduced through the use of polymer in vacuum impermeability of DSP material, as indicated in Table 6.2 (Nazari et al., 2010).

Table 6.2. *Temperature Resistance of DSP Materials*

Temp. (○C)	**Wt. loss (%)**	**Flexural Strength (psi) before PI**	**Flexural Strength (psi) after PI**	**Vac. Leak Rate (torr/ min) before PI**	**Vac. Leak Rate (torr/ min) after PI**
200	4.5	4,650	4,850	0.1	n/a
300	5.0	4,200	4,900	6.0	0
400	6.2	3,850	6,550	11.0	0
550	6.6	3,500	6,250	18.0	0.3
700	7.0	2,450	12,850	100.0	n/a

Source: Roberto A Creative Commons License.

6.3. ENGINEERED CEMENTITIOUS COMPOSITE

Learning Objectives

- Learn the key factors and methods for achieving high-strength concrete.
- Understand the properties and advantages of MS and DSP materials.
- Explore the applications and considerations of DSP materials in construction.

Distinctive variations of cement-based composites are constituted by Engineered cementitious composites (ECC), that incorporate random short fibers for reinforcement. ECC is known for its remarkable pseudo-ductility and toughness. Research conducted at the National Science Foundation Institute the Northwestern University in the United States has developed new cement-based materials that demonstrate substantial enhancements in ductility. These materials are made utilizing the extrusion method and include a tiny proportion of polyvinyl alcohol fiber. The research performed at HKUST, Hong Kong, showed that extruded plates can experience substantial deflection if subjected to four-point bending.

The development of ECC was based on groundbreaking research conducted by the IPC group. Fracture mechanics ideas were employed to analyze structures of fiber-reinforced cementitious composites, a theoretical foundation for its advancement. The microstructure for the composite in ECC was optimized using micromechanics, considering the mechanical relationships among the fiber, matrix, and interface. In a study conducted by Marshall and Cox in 1988, it was demonstrated that under certain conditions, steady-state crack propagation can occur instead of the more commonly observed Griffith-type crack. This phenomenon was found to be governed by a specific equation (Yu et al., 2018):

$$J_{\text{tip}} = \sigma_{\text{ss}}\delta_{\text{ss}} - \int_0^{\delta_{ss}} \sigma(\delta)\, d\delta$$

It is crucial to take into account several factors, including the energy at the crack tip, the ongoing ambient tensile stress, the steady crack opening, and the closing stress to be an outcome of a crack opening. J_{tip} investigates the matrix durability $\frac{K_m^2}{E_m}$ at a low fiber count. Various factors, such as the mix design, water-to-cement ratio, and sand quantity, can have an impact on the strength of fracture K along with Young's modulus E_m (Wu et al., 2012). The energy provided by external work can be conceptualized as the difference between the total energy and the energy lost due to deformation in inelastic springs within the fracture process zone. The idea of the inelastic spring offers a practical approach to comprehending the inelastic occurrences that take place throughout fiber deformation, fiber breaking, and interface breaking down or sliding in the crack zone. Hence, the equation presented above describes the energy equilibrium that takes place during the continuous advancement of fractures in a stable state (Singh et al., 2019).

Figure 6.13 presents a schematic illustration of the power balance concept, demonstrating a fiber that connects the correlation between stress and crack openings. The area A on the right side of the formula is often called the complementary energy. Considering that the maximum value (area B) over this energy J_b' is achieved when σ attains the highest stress σ_0, or δ reaches its crack opening δ_0, this implies a limitation upon the matrix durability as a prerequisite for strain hardening (Shoji et al., 2022):

$$\frac{K_m^2}{E_m} \leq \sigma_0\delta_0 - \int_0^{\delta_0} \sigma(\delta)\, d\delta \equiv J_b'$$

From the equation provided, it becomes evident that the effective design of an ECC necessitates the careful adjustment of fiber, matrix, and interface properties. It is crucial to consider the influence of fiber and interface characteristics on the form of the $\sigma - \delta$ curve. These factors play a significant role in determining J_b.' Optimizing the fiber/matrix interface is crucial in achieving the maximum J_b' value for strain hardening in composite design. It is crucial to carefully consider the composition of the matrix. One important factor to remember is ensuring the fact that a value of J = $K^2{}_m / E_m$ is within an acceptable range (Lepech & Li, 2009).

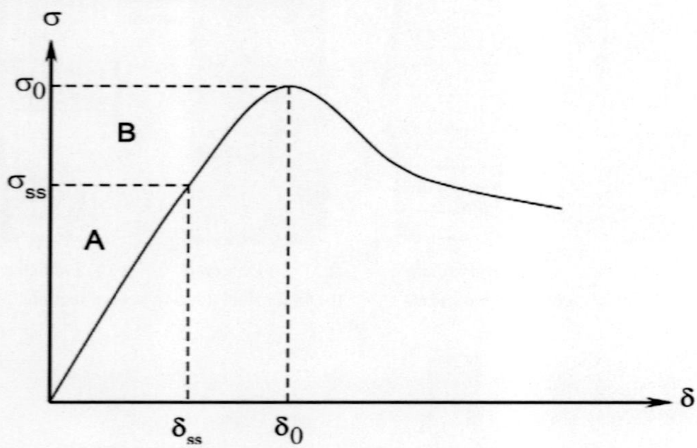

Figure 6.13. *Schematic illustration of energy balance concept on a fiber bridging stress-crack opening stress-strain curve.*

Source: Sanjay Kumar, Creative Commons License.

6.4. FLY ASH CONCRETE OF HIGH-VOLUME

Learning Objectives

- Understand the environmental benefits and drawbacks of incorporating fly ash into concrete.

The material known as fly ash has become a byproduct of a power plant that utilizes coal as its source of fuel. It was previously discarded as industrial waste. Due to the significant annual production of fly ash by power plants, the disposal of fly ash has emerged as a prominent environmental concern in numerous countries. Since the start of the 1960s, numerous countries have taken steps to address the issue of waste disposal by incorporating fly ash through concrete (Siddique, 2004).

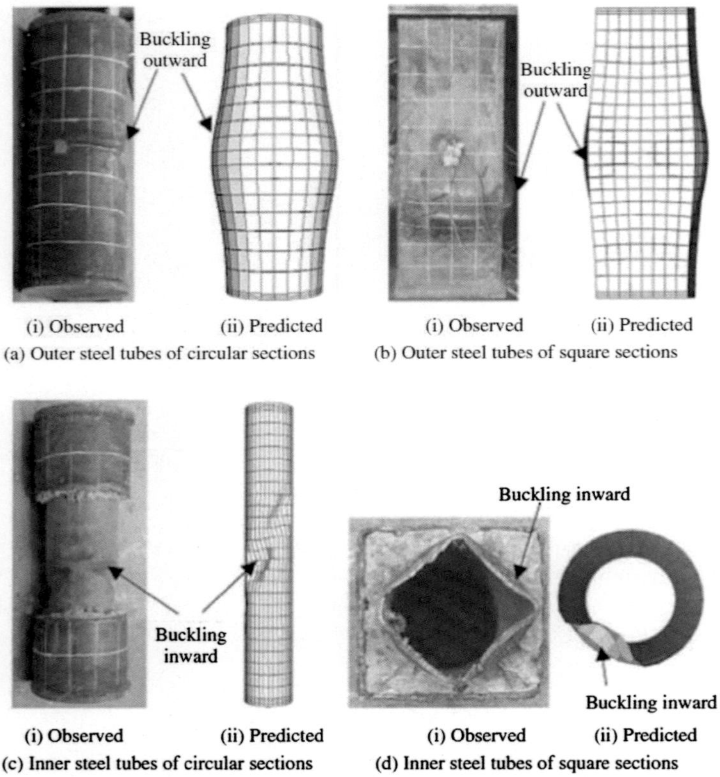

Figure 6.14. *Representation of typical failure modes of hollow CFST.*

Source: Huang ET. Creative Commons License.

Research conducted over the last 50 years has revealed both the advantages and disadvantages of incorporating fly ash into cementitious composites. Fly ash concrete offers significant benefits such as cost-effectiveness, reduced environmental impact, lower heat generation during hydration, and enhanced long-term durability and strength. There are certain drawbacks associated with fly ash concrete, such as a longer time for the initial setting and lower strength in the early stages. The concrete's composition will vary based on the quantity of fly ash used (Atis, 2003).

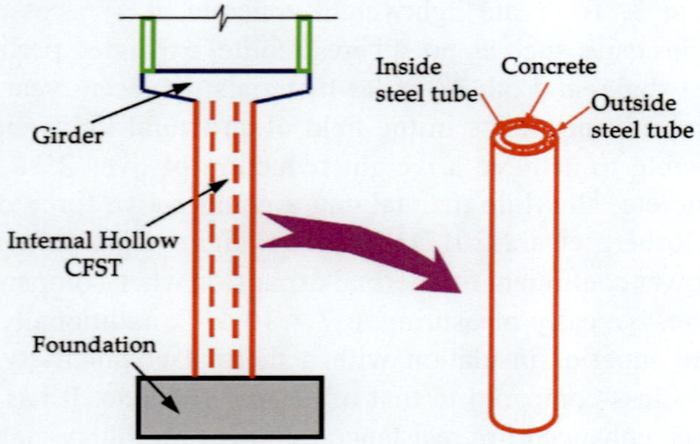

Figure 6.15. *Illustration of composite-type CFST tall bridge pier.*

Source: Johnson K. Creative Commons License.

Fly ash concrete constitutes two categories: regular high-volume fly ash and concrete fly ash concrete. Typically, fly ash concrete incorporates approximately 25–30% of fly ash about the weight of the cementitious binder. This type of concrete is commonly employed in the construction of large-scale frameworks, like gravity dams. Many countries have also made efforts to incorporate a significant quantity of fly ash in concrete while maintaining its early-age compressive strength (Reiner & Rens, 2006). High-volume fly ash (HVFA) building materials can make up 60% or more of the weight of cement with fly ash. By carefully choosing the mix proportions and incorporating chemical admixtures like superplasticizers, HVFA concrete has been able to attain excellent mechanical and durability attributes (Dragaš et al., 2014).

6.5. LIGHTWEIGHT AND HEAVYWEIGHT STRUCTURAL CONCRETE

Learning Objectives

- Recognize the characteristics and applications of structural lightweight and heavyweight concrete.

Structure lightweight concrete is characterized by its high strength at compression, exceeding 17 MPa, and its low bulk density, which is less than 1950 kg/m^3. To create lightweight concrete, it is necessary to utilize lightweight aggregates such as no sphere, zeolite, expanded perlite, expanded clay, expanded shale, and other similar materials. In recent years, there have been significant advancements in the field of structural lightweight concrete. It is now possible to achieve a weight reduction of over 25% compared to traditional concrete, all while maintaining a compressive force of 60 MPa or even higher (Roberz et al., 2017). Structural lightweight cement exhibits a significantly lower coefficient for thermal expansion when compared to normal-weight concrete, typically measuring at 7×10^{-6}/°C. additionally, lightweight concrete offers superior insulation with a thermal conductivity that can be more than 25% less compared to that of normal concrete. It has been widely reported that the enhanced fire resistance for structural lightweight concrete is due to its smaller coefficient of expansion under heat and minimal reduction of strength at high temperatures. Structural lightweight concrete has been widely used in both infrastructure and building construction for over a century. High-performance lightweight concrete has been widely utilized in numerous bridges and tall buildings across the US, Europe, and China. Notably, certain types of lightweight concrete have demonstrated improved durability in terms of chemical resistance, frost resistance, and reduced permeability (Thienel et al., 2020). Heavyweight concrete is typically characterized by its high bulk density, falling within an acceptable range of 3360–3840 kg/m^3. This type of concrete is specifically designed to serve unique and specialized purposes. Heavyweight concrete has been utilized to safeguard against seepage from radioactive structures, as the detrimental impact of radioactive rays (such as x-rays, gamma rays, and nuclear radiation) upon living organisms is well-documented. It finds application in laboratories, healthcare facilities, and thermal power plants. One crucial aspect to consider when dealing with heavyweight building materials is calculating the water-to-cement ratio. An optimal water-to-cement ratio of 0.40 is commonly used in heavyweight concrete. Another crucial consideration for heavyweight building materials is ensuring the right balance of cement dosage. It should be sufficient to ensure impermeability to radiation, while also being kept sufficiently low to prevent cracks caused by shrinkage. It is advisable to use a cement dosage of more than 350 kg/m^3 (Nadesan & Dinakar, 2017).

Specialized heavyweight building materials can also serve as an effective soundproofing solution. Within this particular form of building materials, the aggregates can consist of dense metal balls that are enveloped in a layer of supple rubber, as depicted in Figure 6.16. Indeed, these aggregates can function as a localized resonance unit, effectively blocking sound at specific frequencies. Figure 6.16 displays the experimental findings regarding sound shielding using various construction materials, such as concrete plates, gypsum boards, or plastics. The sound loss in transmission for heavyweight concrete is significantly higher at around 180 Hz, measuring 45 dB, compared to the other materials which only reach 25 dB. It can be inferred that the sound shielding capacity of heavyweight concrete at a rate of 180 Hz is significantly superior to that of other construction materials, with a difference of 100-fold (Khalaf et al., 2019).

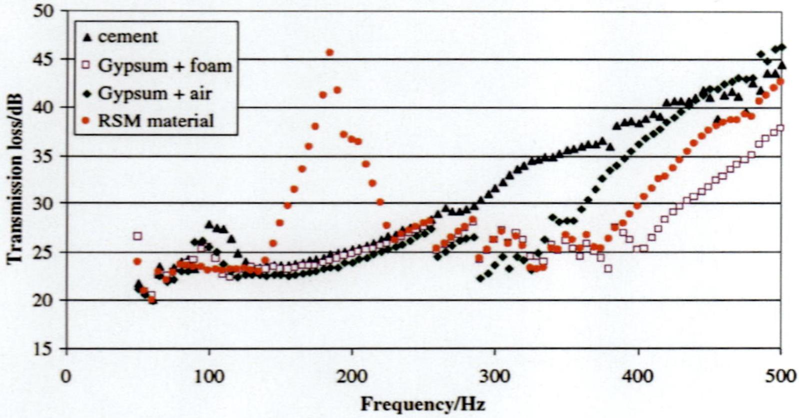

Figure 6.16. *Experimental results of sound shielding with different construction materials.*

Source: Huang ET. Creative Commons License.

Practice Problem

SAMO Constructions is a leading firm specializing in comprehensive construction services, renowned for delivering high-quality projects with a commitment to excellence and client satisfaction. SAMO needs to design a structural element for a high-rise building that requires both lightweight and heavyweight properties, posing a challenge in balancing strength and durability with weight reduction.

Solutions to Practice Problem

SAMO Constructions can optimize the concrete mix design by incorporating a blend of lightweight and heavyweight aggregates, along with supplementary cementitious materials and advanced chemical admixtures. Rigorous testing and quality control measures should be implemented to ensure consistency and reliability of concrete properties throughout production. This approach enables SAMO Construction Firm to tailor Structural Lightweight and Heavyweight Concrete to meet the project's specific demands, achieving the desired balance of properties effectively.

SUMMARY

- The chapter on advanced cementitious composites explores innovative materials and techniques to improve traditional concrete. It begins by analyzing Fiber-Reinforced Cementitious Composites (FRCC), emphasizing factors like fiber type, volume fraction, and processing methods that influence their properties, enhancing mechanical strength and bond properties.
- High-strength cementitious Composites, including high-strength concrete and DSP materials, are discussed for their superior strength and durability in modern construction.
- Engineered Cementitious Composite (ECC) is highlighted for its exceptional ductility and crack control, promising resilient infrastructure solutions.
- High-volume fly Ash Concrete is introduced as an eco-friendly option, reducing environmental impact while enhancing performance.
- Structural Lightweight and Heavyweight Concrete are compared for their unique applications, with lightweight concrete offering reduced dead load and improved insulation, and heavyweight concrete meeting specialized project needs.
- Overall, these advancements pave the way for sustainable, resilient, and high-performance construction solutions in contemporary infrastructure projects.

MULTIPLE CHOICE QUESTIONS

1. Which factor primarily influences the properties of Fiber-Reinforced Cementitious Composites?

a. Matrix Variation

b. Fiber Type

c. Processing Methods

d. Fiber Volume Fraction

2. What aspect of Fiber-Reinforced Cementitious Composites does Fiber Volume Fraction primarily affect?

a. Mechanical Properties

b. Fiber-Cement Bond Properties

c. Matrix Variation

d. Processing Methods

3. Which of the following is not a factor influencing the properties of Fiber-Reinforced Cementitious Composites?

a. Fiber Type

b. Fiber Volume Fraction

c. Aggregate Size

d. Matrix Variation

4. What aspect of Fiber-Reinforced Cementitious Composites is affected by Matrix Variation?

a. Fiber Type

b. Fiber Volume Fraction

c. Mechanical Properties

d. Processing Methods

5. Which type of cementitious composite typically exhibits higher strength compared to conventional concrete?

a. High-volume fly Ash Concrete

b. Engineered Cementitious Composite

c. High-Strength Concrete

d. Structural Lightweight Concrete

6. What does DSP stand for in the context of cementitious composites?

a. Dense Slurry Precast

b. Dual Strength Polymer

c. Dispersion Strengthened Particles

d. Durable Structural Panel

REVIEW QUESTIONS

1. What factors influence Fiber-Reinforced Cementitious Composites' (FRCC) properties?
2. Name two types of High-Strength Cementitious Composites discussed and their benefits.
3. What distinguishes Engineered Cementitious Composite (ECC) from traditional concrete?
4. How does High-Volume Fly Ash Concrete contribute to sustainability?
5. What are the advantages of Structural Lightweight Concrete?

6. When would Structural Heavyweight Concrete be preferred in construction?
7. What are the potential challenges in adopting advanced cementitious composites?
8. How do advanced cementitious composites contribute to resilient infrastructure?

Answers to Multiple Questions

1. (b) 2. (a) 3. (c) 4. (c) 5. (c) 6. (a)

REFERENCES

1. Abdallah, S., Fan, M., & Rees, D. W. (2018). Bonding mechanisms and strength of steel fiber–reinforced cementitious composites: Overview. *Journal of Materials in Civil Engineering, 30*(3), 04018001. https://doi.org/10.1061/(ASCE)MT.1943-5533.0002351.
2. Atıs, C. D. (2003). High-volume fly ash concrete with high strength and low drying shrinkage. *Journal of Materials in Civil Engineering, 15*(2), 153–156. https://doi.org/10.1061/(ASCE)0899-1561(2003)15:2(153).
3. Beglarigale, A., Aydın, S., & Kızılırmak, C. (2016). Fiber-matrix bond characteristics of alkali-activated slag cement–based composites. *Journal of Materials in Civil Engineering, 28*(11), 04016133. https://doi.org/10.1061/(ASCE)MT.1943-5533.0001718.
4. Cuenca, E., & Ferrara, L. (2017). Self-healing capacity of fiber reinforced cementitious composites. State of the art and perspectives. *KSCE Journal of Civil Engineering, 21*, 2777–2789. https://doi.org/10.1007/s12205-017-0358-8.
5. de Andrade Silva, F., Mobasher, B., Soranakom, C., & Toledo Filho, R. D. (2011). Effect of fiber shape and morphology on interfacial bond and cracking behaviors of sisal fiber cement-based composites. *Cement and Concrete Composites, 33*(8), 814–823. https://doi.org/10.1016/j.cemconcomp.2011.05.009.
6. Dhakal, H. N., Zhang, Z. A., & Richardson, M. O. (2007). Effect of water absorption on the mechanical properties of hemp fiber reinforced unsaturated polyester composites. *Composites Science and Technology, 67*(7–8), 1674–1683. https://doi.org/10.1016/j.compscitech.2006.08.022.
7. Ferreira, S. R., de Andrade Silva, F., Lima, P. R. L., & Toledo Filho, R. D. (2015). Effect of fiber treatments on the sisal fiber properties and fiber-matrix bond in cement-based systems. *Construction and Building Materials, 101*, 730–740. https://doi.org/10.1016/j.conbuildmat.2015.10.095.

8. Guerrini, G. L. (2000). Applications of high-performance fiber-reinforced cement-based composites. *Applied Composite Materials, 7*, 195–207. https://doi.org/10.1023/A:1008262519638.

9. He, J., Watkins, S., & Kelley, D. E. (2001). Skeletal muscle lipid content and oxidative enzyme activity in relation to muscle fiber type in type 2 diabetes and obesity. *Diabetes, 50*(4), 817–823. https://doi.org/10.2337/diabetes.50.4.817.

10. Hikida, R. S., Staron, R. S., Hagerman, F. C., Walsh, S., Kaiser, E., Shell, S., & Hervey, S. (2000). Effects of high-intensity resistance training on untrained older men. II. Muscle fiber characteristics and nucleo-cytoplasmic relationships. *The Journals of Gerontology Series A: Biological Sciences and Medical Sciences, 55*(7), 347–354. https://doi.org/10.1093/gerona/55.7.B347.

11. Huang, J. S., Liew, J. X., & Liew, K. M. (2021). Data-driven machine learning approach for exploring and assessing mechanical properties of carbon nanotube-reinforced cement composites. *Composite Structures, 267*, 113917. https://doi.org/10.1016/j.compstruct.2021.113917.

12. Iaquinto, A. L., & Fredrickson, J. W. (1997). Top management team agreement about the strategic decision process: A test of some of its determinants and consequences. *Strategic Management Journal, 18*(1), 63–75. https://doi.org/10.1002/(SICI)1097-0266(199701)18:1<63::AID-SMJ837>3.0.CO;2-L.

13. Jamal Shannag, M., & Hansen, W. (2000). Tensile properties of fiber-reinforced very high strength DSP mortar. *Magazine of Concrete Research, 52*(2), 101–108. https://doi.org/10.1680/macr.2000.52.2.101.

14. Joo Kim, D., Naaman, A. E., & El-Tawil, S. (2008). Comparative flexural behavior of four fiber reinforced cementitious composites. *Cement and Concrete Composites, 30*(10), 917–928. https://doi.org/10.1016/j.cemconcomp.2008.05.002.

15. Kanakubo, T. (2006). Tensile characteristics evaluation method for ductile fiber-reinforced cementitious composites. *Journal of Advanced Concrete Technology, 4*(1), 3–17. https://doi.org/10.3151/jact.4.3.

16. Khalaf, M. A., Ban, C. C., & Ramli, M. (2019). The constituents, properties and application of heavyweight concrete: A review. *Construction and Building Materials, 215*, 73–89. https://doi.org/10.1016/j.conbuildmat.2019.04.152.

17. Khan, M. I. (2012). Evaluation of non-destructive testing of high strength concrete incorporating supplementary cementitious composites. *Resources, Conservation and Recycling, 61*, 125–129. https://doi.org/10.1016/j.resconrec.2011.11.013.

18. Lei, D. Y., Guo, L. P., Li, Y., Zheng, Z., Liu, J. P., Li, S. C., & Zhong, B. M. (2021). The investigating on mechanical properties of ultra-high strength and ultra-high ductility cementitious composites (UHS-UHDCC). *Journal of Building Engineering, 43*, 102486. https://doi.org/10.1016/j.jobe.2021.102486.
19. Lepech, M. D., & Li, V. C. (2009). Water permeability of engineered cementitious composites. *Cement and Concrete Composites, 31*(10), 744–753. https://doi.org/10.1016/j.cemconcomp.2009.07.004.
20. Li, V. C. (2003). On engineered cementitious composites (ECC): A review of the material and its applications. *Journal of Advanced Concrete Technology, 1*(3), 215–230. https://doi.org/10.3151/jact.1.215.
21. Lima, P. R., Santos, H. M., Camilloto, G. P., & Cruz, R. S. (2017). Effect of surface biopolymeric treatment on sisal fiber properties and fiber-cement bond. *Journal of Engineered Fibers and Fabrics, 12*(2), 155–165. https://doi.org/10.1177/155892501701200210.
22. Mir, N. A., Rafiq, A., Kumar, F., Singh, V., & Shukla, V. (2017). Determinants of broiler chicken meat quality and factors affecting them: A review. *Journal of Food Science and Technology, 54*(11), 2997–3009. https://doi.org/10.1007/s11483-017-0514-8.
23. Mobasher, B., & Li, C. Y. (1996). Effect of interfacial properties on the crack propagation in cementitious composites. *Advanced Cement Based Materials, 4*(3–4), 93–105. https://doi.org/10.1016/S1065-7355(96)90002-7.
24. Nadesan, M. S., & Dinakar, P. (2017). Structural concrete using sintered fly ash lightweight aggregate: A review. *Construction and Building Materials, 154*, 928–944. https://doi.org/10.1016/j.conbuildmat.2017.08.074.
25. Pacheco, F., Christ, R., Quinino, U., & Tutikian, B. F. (2018). Effects of fiber hybridization in advanced cementitious composites durability in humid and aggressive environments. *Matéria (Rio de Janeiro), 23*, 12171. https://doi.org/10.1590/S1517-707620180001.0786.
26. Pan, G., Sun, W., Ding, D., & Zhang, Y. (1998). Experimental study on the micro-aggregate effect in high-strength and super-high-strength cementitious composites. *Cement and Concrete Research, 28*(2), 171–176. https://doi.org/10.1016/S0008-8846(98)00009-1.
27. Prabha, S. L., Gopalakrishnan, M., & Neelamegam, M. (2020). Development of high-strength nano-cementitious composites using copper slag. *ACI Materials Journal, 117*(4), 37–46. https://doi.org/10.14359/51729577.
28. Prakash, S., & Kumar, P. (2023). Environmentally friendly ultra-high-performance concrete enhanced with supplementary cementitious materials. *Building a Blueprint for Progress, 28*, 98–100.

29. Ranade, R., Li, V. C., Stults, M. D., Heard, W. F., & Rushing, T. S. (2013). Composite properties of high-strength, high-ductility concrete. *ACI Materials Journal, 110*(4), 987–990. https://doi.org/10.14359/51634657.

30. Ravitheja, A., Kumar, G. P., & Anjaneyulu, C. M. (2021). Impact on cementitious materials on high strength concrete–A review. *Materials Today: Proceedings, 46*, 21–23. https://doi.org/10.1016/j.matpr.2020.08.289.

31. Reiner, M., & Rens, K. (2006). High-volume fly ash concrete: Analysis and application. *Practice Periodical on Structural Design and Construction, 11*(1), 58–64. https://doi.org/10.1061/(ASCE)1084-0680(2006)11:1(58).

32. Roberz, F., Loonen, R. C. G. M., Hoes, P., & Hensen, J. L. M. (2017). Ultra-lightweight concrete: Energy and comfort performance evaluation in relation to buildings with low and high thermal mass. *Energy and Buildings, 138*, 432–442. https://doi.org/10.1016/j.enbuild.2016.12.039.

33. Sahmaran, M., Lachemi, M., & Li, V. C. (2010). Assessing mechanical properties and microstructure of fire-damaged engineered cementitious composites. *ACI Materials Journal, 107*(3), 631–635. https://doi.org/10.14359/51606546.

34. Savastano Jr, H., Santos, S. F. D., Radonjic, M., & Soboyejo, W. O. (2009). Fracture and fatigue of natural fiber-reinforced cementitious composites. *Cement and Concrete Composites, 31*(4), 232–243. https://doi.org/10.1016/j.cemconcomp.2009.01.005.

35. Shah, S. P., & Li, Z. (2003). Bond properties in fiber-cement and aggregate-cement paste interfaces. In *Fracture Mechanics of Concrete Structures* (Vol. 2, pp. 797–808). CRC Press.

36. Shoji, D., He, Z., Zhang, D., & Li, V. C. (2022). The greening of engineered cementitious composites (ECC): A review. *Construction and Building Materials, 327*, 126701. https://doi.org/10.1016/j.conbuildmat.2022.126701.

37. Siddique, R. (2004). Performance characteristics of high-volume Class F fly ash concrete. *Cement and Concrete Research, 34*(3), 487–493. https://doi.org/10.1016/j.cemconres.2003.10.004.

38. Song, H., Liu, J., He, K., & Ahmad, W. (2021). A comprehensive overview of jute fiber reinforced cementitious composites. *Case Studies in Construction Materials, 15*, e00724. https://doi.org/10.1016/j.cscm.2021.e00724.

39. Thienel, K. C., Haller, T., & Beuntner, N. (2020). Lightweight concrete—From basics to innovations. *Materials, 13*(5), 1120. https://doi.org/10.3390/ma13051120.

40. Tran, T. K., & Kim, D. J. (2014). High strain rate effects on direct tensile behavior of high-performance fiber reinforced cementitious composites. *Cement and Concrete Composites, 45*, 186–200. https://doi.org/10.1016/j.cemconcomp.2013.09.003.

41. Wu, M., Johannesson, B., & Geiker, M. (2012). A review: Self-healing in cementitious materials and engineered cementitious composite as a self-healing material. *Construction and Building Materials, 28*(1), 571–583. https://doi.org/10.1016/j.conbuildmat.2011.10.013.

42. Yang, E. H., Sahmaran, M., Yang, Y., & Li, V. C. (2009). Rheological control in production of engineered cementitious composites. *ACI Materials Journal, 106*(4), 357–364. https://doi.org/10.14359/0000000000000007.

43. Yoo, D. Y., Lee, J. H., & Yoon, Y. S. (2013). Effect of fiber content on mechanical and fracture properties of ultra high-performance fiber reinforced cementitious composites. *Composite Structures, 106*, 742–753. https://doi.org/10.1016/j.compstruct.2013.06.017.

44. Yu, K., Li, L., Yu, J., Wang, Y., Ye, J., & Xu, Q. (2018). Direct tensile properties of engineered cementitious composites: A review. *Construction and Building Materials, 165*, 346–362. https://doi.org/10.1016/j.conbuildmat.2018.01.111.

CHAPTER 7

FIBER REINFORCED COMPOSITES IN CIVIL INFRASTRUCTURE

LEARNING OBJECTIVES

At the end of this chapter, readers will be able to:

1. Understand the composition and functionality of FRP composites in construction.
2. Know the advantages of FRP composites over conventional construction materials.
3. Know the applications of FRP composites in civil infrastructure rehabilitation.
4. Understand the potential impact of FRP composites on construction timelines and costs.
5. Identify the challenges in adopting FRP composites in construction.
6. Understand the implications of higher initial material costs and lack of familiarity.
7. Recognize the need for integrated materials-process-design structures in FRP composites.
8. Evaluate the applications of FRP composites in structural rehabilitation and new construction.

INTRODUCTORY EXAMPLE

Structural Solutions Inc. is a leading construction firm specializing in innovative materials and techniques for civil infrastructure projects. Recently, they were tasked with the rehabilitation of an aging bridge in a busy urban area. The bridge, constructed decades ago using traditional materials, was showing signs of deterioration due to corrosion and fatigue. To address these issues and ensure the longevity of the bridge, Structural Solutions Inc. proposed the use of FRP By incorporating FRP materials into various components of the bridge, they aimed to enhance its structural integrity while minimizing maintenance requirements and downtime.

Utilizing FRP composites, Structural Solutions Inc. implemented a comprehensive rehabilitation plan that included reinforcing the bridge deck, strengthening the support columns, and retrofitting critical connections. This approach not only extended the lifespan of the bridge but also improved its performance under various loading conditions, including seismic events. Through their expertise and dedication to innovation, Structural Solutions Inc. successfully demonstrated the efficacy of FRP composites in civil infrastructure projects, paving the way for their wider adoption in the construction industry.

UNIT INTRODUCTION

The combination of reinforcements and binder substances has been widely utilized by designers to create highly efficient construction materials. From the ancient Egyptians using straw as reinforcement within clay bricks and mud, to the innovation of plywood to create a laminated composite, the development of composite material systems has been driven by the demands of building along with transportation frameworks. As an illustration, the fusion of steel reinforcement and concrete has served as the foundation for various structural systems employed in construction throughout the twentieth century (Hollaway, 2003). Researchers and scholars in the field of civil engineering are constantly exploring and embracing innovative materials to enhance the construction of robust, spacious, long-lasting, environmentally friendly, and visually pleasing buildings. The civil engineer has always faced an ever-growing number of challenges due to the demands of society and the forces of nature. These challenges necessitate the ongoing improvement of materials and the creation of new material types and systems that can effectively integrate form and function. This is crucial to ensure that the structures constructed meet the current and future requirements of humanity (Karbhari et al., 2003).

In the current era, civil engineers are confronted with a multitude of challenges and opportunities, with one of the most significant being the revitalization of civil infrastructure amenities. The degradation and operational insufficiency of current civil infrastructure poses a major obstacle for nations worldwide. Furthermore, the degradation of materials due to aging, exposure to harsh weather conditions, and unforeseen structural damage are not the only concerns. The demands of modern society, such as increased traffic and the need for more housing and industrial facilities, have escalated significantly in recent decades (Brandt, 2008). This has led to a shift from local to international transport of services and goods. There is an increasing demand to expand highway systems to handle larger traffic volumes and reinforce existing structures to support heavier loads at higher speeds. Improved comprehension of the building response and the reduction of damage caused by natural disasters like storms and earthquakes have resulted in the development of updated design regulations and the subsequent requirement to restore existing structures for ongoing safety (Fang et al., 2019).

Traditional materials like timber, steel, and concrete offer several benefits, including the affordability of their raw materials. Nevertheless, it is evident that traditional materials and technologies, while adequate for various construction purposes and having a reasonably successful track record, fall short in terms of durability for challenging applications. In certain instances, these are prone to rapid deterioration, underscoring the necessity for advanced construction materials and structures with superior performance.

The application of FRP or fiber-reinforced polymer, matrix composites in civil infrastructure holds great promise, offering opportunities for the revitalization of current structures and the creation of new facilities. Originally designed for defense and aerospace sectors, these materials have proven to be highly versatile. This chapter aims to provide an overview of the current state of emerging building products in civil infrastructure. Instead of providing an overall description of FRP composites, this article focuses on their specific application in civil construction. An analysis is conducted on the factors that impact the advancement of composites just like structural materials for both rehabilitation and new building projects, taking into consideration their specific applications (Uomoto et al., 2002).

7.1. FIBER REINFORCED COMPOSITES AS A CLASS OF CONSTRUCTION MATERIALS

Learning Objectives

- Understand the composition and advantages of fiber-reinforced composites.
- Know the potential of composites to address civil infrastructure demands.

7.1.1. An Emerging Class of Construction Materials

This chapter focuses on a specific definition of fiber-reinforced composites. These composites consist of polymeric resins that serve as matrices or binders, combined with durable and rigid fiber assemblies that function as the phase of reinforcement. The incorporation of a matrix phase alongside a reinforcing phase, results in the creation of a novel material system that can be compared to steel-concrete reinforced in terms of its concept, despite the significant differences in the proportions of the reinforcing fractions.

In general, reinforced concrete typically has a low percentage of reinforcement, usually less than 5%. However, in the case of fiber-reinforced composite materials, the volume fraction of reinforcement can range from 35% to 65%. Unlike concrete because the matrix of these composites can withstand both tension and compression. Additionally, the polymeric resins utilized in this specific area contribute to the overall toughness of the FRP composite (Bobovich, 2019).

This discussion will focus solely on the performance characteristics of FRP composites, specifically their mechanical, chemical, physical, thermal energy, and electromagnetic characteristics. The strength and stiffness characteristics of the fibers are crucial, while the matrix plays a role in load transfer and protecting the environment. Commonly used materials for fiber reinforcement include carbon, glass, and aramid. Commonly utilized thermoset polymer matrices for composite materials in infrastructure include polyesters, epoxies, vinyl esters, and phenolic (Kromoser et al., 2019).

FRP composites have gained popularity as construction materials due to their unique advantages. These materials offer a high level of customization by combining different types of fibers with a polymeric resin matrix. The fibers are strategically placed to carry load in specific directions, taking advantage of their anisotropic properties. Meanwhile, the resin not only facilitates stress transfer between adjacent fibers through adhesion but also offers security for the fibers. The wide range of material choices available to designers allows them to tailor the selection to meet the specific needs of the structure. This presents a great opportunity to strengthen the existing materials used in civil infrastructure (Khalid et al., 2021).

7.1.2. Solutions for Civil Infrastructure Demands

When exploring material systems for civil infrastructure, fiber-reinforced composites frequently emerge as a viable option. These materials provide a robust, long-lasting, and lightweight solution that is also easy to install, making them a compelling alternative to traditional construction methods. The findings from various research projects on composites, which were launched almost ten years ago in different countries including the U.S., Europe, Japan, and Canada, are currently being integrated into multiple infrastructure developments. The utilization of composites in infrastructure has resulted in notable progress in the creation of affordable reinforcing systems for concrete both internal and external reinforcement, as well as pre-stressing cables and tendons, durable materials. Additionally, it has led to the creation of durable materials, dependable joining techniques, optimized components for structure, and efficient repair techniques (Ferrara, 2019).

The rise of fiber-reinforced composites in the construction industry can be attributed to the need for alternative materials to address the issues of corrosion in steel reinforcement. These composites have found applications in both the internal and external positive reinforcement of concrete, as well as in pre-stressing cables and tendons. In addition, composites have proven to be effective in enhancing the strength of wood materials (Belarbi et al., 2016). In the construction equipment market, competition often requires the use of materials that are both cost-effective and durable, requiring minimal maintenance over a long period. Composites used in construction vary from those commonly used in aerospace applications in terms of their material constituents, manufacturing processes, and joining methods, to achieve these objectives. An interesting application involves the use of multi-axial heavy stitched E-glass textiles for reinforcing large structural components made through the pultrusion method or vacuum-assisted resin transfer molding (VARTM) (Naser et al., 2019).

7.2. ADVANTAGES OF FRP COMPOSITES OVER CONVENTIONAL CONSTRUCTION MATERIALS

Learning Objectives

- Recognize the high specific strength and stiffness advantages of fiber reinforced composites.
- Understand the enhanced fatigue life and its implications for structural design.
- Know the benefits of corrosion resistance and non-magnetic properties in composites.
- Appreciate the potential for lower life-cycle costs and faster field installation with composites.

Outlined below are several significant advantages that are thoroughly examined within the realm of civil infrastructure applications:

7.2.1. High Specific Strength and Stiffness

The strength-to-weight ratios of composites made from fibers exhibit significant enhancements. As an illustration, Table 7.1 provides a comparison between the typical ranges of characteristics for FRP composites (specifically glass and fiberglass reinforced) and reinforcing steel. FRP can demonstrate superior stiffness-to-weight ratios when compared to metals and alloys, as illustrated in Table 7.1. The remarkable combination of exceptional strength and stiffness provides designers with a valuable advantage, allowing them to create designs with reduced weights and thicknesses. In addition, these attributes allow civil engineers to explore novel design ideas, such as bridges with extended free spans, which would be constrained by the unique properties of alternative construction materials (Das & Nizam, 2014).

Table 7.1. *Comparison of Typical Range of FRP Composite Properties with those of Mild Steel*

Property	**Range**	**Comparison with Steel**
Modulus	20 - 138 GPa	1/10 to 2/3 that of steel
Strength	340 - 1700 MPa	1 to 5 times the yield strength of mild steel
Strain to Failure	1 – 3 %	1/10 to 1/3 that of mild steel
Density	1.4 - 2.0 g/	4 to 6 times lighter than steel

Source: Tarun R. Creative Commons License.

In the field of civil infrastructure, the utilization of fiber-reinforced composites can lead to significant improvements. These include increased seismic resistance, faster construction times, and reduced fabrication durations for large structures (Zaman et al., 2013). In offshore applications, the weight savings within the superstructure lead to significant savings in supporting structure and substructure components. When it comes to bridges, this would result in a reduced requirement for extensive foundations. Similarly, in buildings, it would greatly diminish the dimensions of columns on lower levels, allowing for more floor space and the possibility of constructing larger halls for concerts and auditoriums without any limitations on the clear span (Gudonis et al., 2013).

7.2.2. Enhanced Fatigue Life

Many composites have a high fatigue resistance, allowing designers to have more flexibility in their structural designs. Carbon/epoxy composites offer a significant advantage over metal frames in the aerospace industry (Vassilopoulos et al., 2008).

In the field of structural materials, researchers often employ an S-N diagram for evaluating fatigue properties. This diagram plots the amplitude of stress or greatest stress against the number of cycles, revealing that as the stress level decreases, the material's resistance to failure increases over time. If fatigue failure fails to occur below a particular stress threshold, it may be assumed that the material has an infinite lifespan. The stress limit is commonly referred to as the fatigue and endurance limitation. An endurance limit has been identified at 105 to 106 cycles for stainless steel and some of the other alloys (Orcesi et al., 2019). In the realm of fiber-reinforced composites, it is often observed that obtaining a clear-cut endurance limit is not always possible. However, it is worth noting that the S-N curve does exhibit a significant reduction in slope at lower stress levels. Typically, in such scenarios, it is customary in design practice to designate the resistance to fatigue of the material made from FRP at extremely high numbers of cycles, such as 106 to 107 cycles, to serve as the power limit. In contrast to metals, FRP composites may experience a gradual decrease in stiffness or softening under cyclic loads, even before any visible cracks become apparent. This phenomenon is attributed to microscopic damage. For instance, in the case of unilateral composites made of fiber-reinforced materials loaded parallel with the fibers, there are several potential fatigue harm mechanisms. These include matrix cracking, fiber breakage, interfacial debonding, and interfacial shear failure (Sharma & Joshi, 2023).

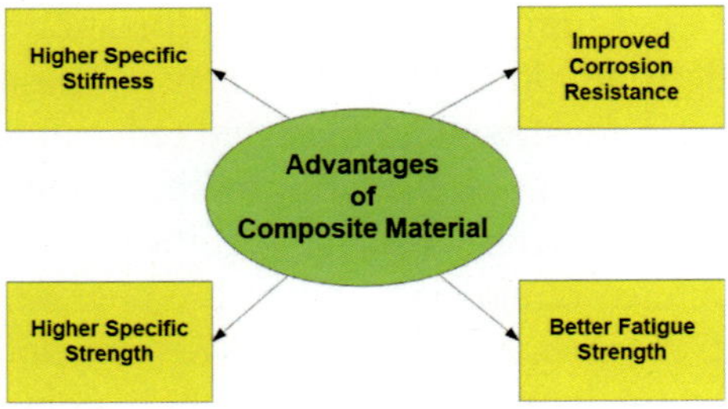

Figure 7.1. *Illustration of the Advantages of composite material over a conventional material.*

Source: Sagar Choski. Creative Commons License.

Fatigue along with fretting fatigue can cause damage and cracking, leading to significant emotional distress within bridge and construction components. Reduced fatigue worries associated with composites can drive substantial advancements in structural design, particularly in earthquake-prone regions or in buildings that accommodate heavy vibrating equipment. For instance, bridge decks constructed using E-glass/vinyl ester composites through pultrusion and vacuum-assisted resin transfer molding (VARTM), as well as FRP-concrete hybrid decks, have demonstrated remarkable durability in fatigue tests, enduring up to two million load cycles without any signs of damage buildup. Nevertheless, the durability of bonded or bolted connections could potentially impact the overall lifespan of the structure. In structural applications that prioritize stiffness, such as FRP bridge decks, the working stresses are usually kept very low in comparison to the ultimate strength. This leads to a longer fatigue life for the composites (Noël, 2019).

7.2.3. Corrosion Resistance

Composites have the advantage of not rusting, which makes them highly desirable for applications where rust is a major concern. There has been a significant surge in the fascination surrounding reinforcement bars and grids made from composites, along with cables used for pre-and post-tensioning, as well as cable stays. In a report published by BIRL/Northwestern University in 1995, corrosion mitigation was highlighted as a significant factor driving the use of composite substances in infrastructure (Kossakowski & Wciślik, 2022). Based on this report, it is recommended to utilize composite materials in situations where traditional metallic components require frequent and costly maintenance because of corrosion and its associated consequences. In the context of bridges, one practical application involves the use of a corrosion-resistant sewer system to redirect the run-off of deicing salt. Composites have found application in various fields for their excellent corrosion resistance. These include pipelines, marine waterfront structures, cladding panels, mining, and walkways in challenging environments (Frigione & Lettieri, 2018).

7.2.4. Non-magnetic Properties

Steel members along with reinforcements can pose challenges in structures like hospital working theaters, areas with radar operations, and spaces housing sensitive electronic devices. This is because they tend to create interference when exposed to electromagnetic waves. Fiberglass-reinforced composite parts are frequently employed as a viable solution to address these issues (Li et al., 2018).

7.2.5. Lower Life-cycle Costs

Composite structures offer superior corrosion resistance and increased durability against solvents and environmental factors. Minimal maintenance is required, leading to significant cost savings over their lifespan. Empirical evidence strongly suggests that, when meticulously engineered and manufactured, structures made of fiber-reinforced composites exhibit remarkable durability, leading to significantly reduced costs over their lifespan. Unfortunately, there is a lack of comprehensive long-term data that is properly documented and readily available to civil engineers. In addition, the data on strength can often be conflicting, with different kinds of material and procedures leading to uncertainty (Ilg et al., 2016).

7.2.6. Faster Field Installation

The construction process in civil engineering is often associated with lengthy installation periods, leading to significant delays in the opening of facilities. These delays can cause inconvenience to users, such as road diversions, and lane blockages, speed limit restrictions, especially during repairs or expansions of existing roads and bridges. In addition, the use of traditional materials in construction is often limited to specific seasons, leading to extended periods of inactivity. On the other hand, the advantage of using large composite parts is that they can be manufactured off-site in manufacturing factories because of their lightweight nature. This allows for easy transportation to the building site and installation using light equipment, which reduces the need for extensive on-site work. The utilization of fiber-reinforced composites in construction can result in the year-round setup of composite structures, leading to improved construction efficiency and positively impacting planning and logistics. Nevertheless, advancements in adhesive bonding techniques for composite structure parts may be necessary to address the challenges posed by different pressure, temperature, and condensation conditions (Wu et al., 2014).

7.3. CHALLENGES IN THE ADOPTION OF FRP COMPOSITES AS CONSTRUCTION MATERIALS

Learning Objectives

- Recognize challenges in cost analysis and manufacturing processes for FRP composites.
- Understand the complexity of material properties and design considerations for composites.
- Understand the need for integration of materials, processes, and design in composite structures.

Despite the numerous benefits and potential of FRP composites, civil engineers need to acknowledge the existing obstacles that need to be addressed before these emerging materials become as widely utilized as traditional materials like timber, concrete, and steel. This section highlights some of the significant challenges. It is important to acknowledge that in numerous instances, these obstacles pertain to abstract concepts that can be overcome through education, research, implementation, and increased project involvement (Siddika et al., 2020).

7.3.1. Higher Initial Material Cost

The significant challenge in incorporating composites into the highly cost-competitive construction market is the perception of higher material costs. The materials used in FRP composites, such as the reinforcement fibers and resin matrix, tend to be pricier compared to traditional construction materials when considering their equivalent mass. Nevertheless, composite materials possess a notable advantage in terms of weight reduction. As explained earlier, they exhibit superior strength-to-weight and stiffness-to-weight ratios, allowing for the same functionality to be achieved with a smaller amount of material. Hence, when evaluating the cost of composites compared to traditional materials, it is crucial to analyze their performance on a per-unit basis, taking into account the potential weight reductions.

In addition to the manufacturing cost, the total expense of the FRP composite must be taken into consideration. These expenses can be decreased by implementing a continuous fabrication method that minimizes the need for manual labor, like pultrusion. On the other hand, there are flexible fabrication methods available for producing large structural components without the need for costly tooling. One such method is vacuum-assisted resin transfer molding (VARTM), which can effectively reduce the overall costs of production (Tao et al., 2023).

7.3.2. Lack of Familiarity with Materials

The elastic and strength characteristics of composite materials can vary significantly based on factors such as the constituents of the material (fiber, matrix, and fillers), the arrangement of the fibers, and the volume fraction of the fibers. To fully describe a composite material with transverse isotropy, it is necessary to consider five independent elastic constants. There exists a significant distinction between fibrous composites and conventional isotropic materials like steel. The latter has achieved standardization by employing accepted grade types, whereas the former lacks such uniformity. The versatility of composites can be both a source of excitement and a challenge for designers. Fiber-reinforced composites can experience various failure mechanisms based on their microstructure and fiber architecture (Mirmiran, 2004). These mechanisms include debonding, fiber breakage, matrix cracking, and fiber pullout. Consequently, designing a structurally sound composite becomes a challenging task. Furthermore, a laminated structure may display areas of vulnerability between layers, along with localized discrepancies in properties caused by resin-rich regions, voids, and fiber misalignments.

However, civil engineers can gain familiarity with composites by studying and implementing requirements and recommendations for material evaluation, which are widely utilized in various industries. These standards encompass a wide range of testing methods, including those for constituent characteristics, thermal and physical characteristics of lamina and laminated material mechanical characteristics, as well as structural properties (Mohammed et al., 2020).

7.3.3. Lack of Knowledge on the Design of Composite Structures

Engineers designing with composites presents a more complex challenge for engineers compared to conventional materials. It involves the simultaneous design of both the material's structure itself is meant along with the symmetry of the part. Engineers who are accustomed to working with traditional construction materials may face challenges when using composites due to the absence of standardized grades and the lack of ability to modify them from stock forms and profiles, which is common alongside steel bars, plates, and profiles. Nowadays, it is necessary to take into account both design and fabrication simultaneously, often by the same person, instead of treating them as separate functions (Cheng & Karbhari, 2006).

However, civil engineers may conquer these challenges by familiarizing themselves with and applying the methodology for analyzing and designing composite structures, as utilized in various other industries. It is necessary to outline the methods used for material analysis and establish their relationship with the adaptable and strength properties that are utilized in the design process. It involves utilizing laminate analysis to forecast hygro-thermal along with mechanical strains and stresses. Utilizing laminate analysis is a pragmatic method for calculating the stiffness characteristics of laminated structural elements. Efficient layup designs can be systematically approached using practical design methods, such as the utilization of carpet plots. Ultimately, a comprehensive grasp of failure modes enables the selection and application of suitable failure criteria for the design of laminate strength (Porter & Harries, 2007).

7.3.4. Need for Integrated Materials-Process-Design Structure

As previously stated, the effectiveness of FRP composites, and the derived components or structures, heavily relies on the seamless integration of materials choice, process technique, and overall design. The combination of these aspects, which are commonly used in traditional construction, is not frequently applied in FRP composites. Consequently, this can pose a significant challenge to gaining acceptance from the civil engineering community (Baldwin et al., 1996). Recent advancements in pultrusion and VARTM processing have led to the integration of materials, procedures, and structural design. This has been accomplished by the use of multiaxial fabric reinforcements and toughened resin matrices. The outcome of these developments is the creation of optimized designs for various structural components such as bridge decks, piles, and linear profiles. However, these advancements necessitate civil engineers to collaborate with professionals from various fields such as composites processing, polymer chemistry, fiber reinforcements, and material characterization (Jacobson et al., 1998).

7.4. MATERIALS PERSPECTIVE FOR FRP COMPOSITES DESIGN

Learning Objectives

- Understand the unique attributes and considerations in designing FRP composites.
- Comprehend the factors influencing material properties and their implications in design.

FRP composites possess distinct qualities as a result of the incorporation of robust and rigid fibers alongside polymeric resin matrices. Consequently, the design of these composites must take into consideration their specific attributes, such as anisotropy, inhomogeneity, layered structure or laminated or Additional factors to consider when designing FRP composites include their non-ductile unidirectional response, various damage mechanisms, and coefficients of thermal expansion, internal damping, and corrosion resistance (Al-Saadi et al., 2018).

It is important to highlight that the characteristics of FRP composites are influenced by the characteristics of the materials used, such as the reinforcing fiber, matrix, and fillers, as well as their respective volume fractions. It is worth noting that while the modulus of a carbon fiber can reach an impressive 230 GPa, surpassing that of Grade 60 metal at 200 GPa, the modulus of a single-directional carbon fiber reinforced combined is significantly lower. Specifically, in the direction of the reinforcement, the modulus is around 160 GPa, and in a direction transverse toward the fiber direction, it drops to just 10 GPa for a fiber volume fraction of 60% (Table 7.2) (Harris, 1991).

Table 7.2. *Properties of Typical Reinforcing Fibers*

Fiber	Designation	Density (g/)	Tensile Modulus (GPa)	Tensile Strength (MPa)	Strain to Failure (%)	Tow Count
Glass	E-Glass	2.55	72.4	3450	4.8	---
	S-Glass	2.49	86.9	4800	5.2	---
Pan Carbon	T-300a,b	1.76	230	3530	1.5	1, 3, 6, 12 K
	AS4c	1.80	221	3930	1.7	3, 6, 12 K
	IM6c	1.74	275	5240	1.7	12 K
	UHMc	1.85	441	3445	0.8	3, 6, 12 K
Pitch Carbon	P-55a	2.0	380	1900	0.5	---
	P-100a	2.15	758	2410	0.32	---
Aramid	Kevlar 49d	1.45	131	3620	2.8	---
	Technorag	1.39	74	3500	4.6	---

Source: Lopez Anido, Creative Commons License.

When selecting the appropriate reinforcement and polymer matrix, it is crucial to consider factors such as the intended structural use, the environment around it, the desired lifetime, and the chosen manufacturing process. In the field of civil infrastructure, there is often a preference for one type of fiber over another, such as carbon over glass or glass of carbon. However, it is important to note that both of them have their own set of pros and cons. Therefore, the selection of fiber type should be based on the specific details and necessities of the application being considered (Zhou et al., 2022).

Carbon fibers come in various moduli, with certain grades reaching moduli that are 3 to 4 times higher than steel. Nevertheless, in all these different types, the strain capacity is significantly restricted. On the other hand, glass fibers possess a considerably higher strain capacity, making them more suitable for designs that involve non-catastrophic failure methods. Carbon fibers have a chemical structure that makes them highly resistant to environmental factors. This quality allows them to be utilized safely in programs that involve exposure to harsh environmental conditions. The environmental impact of glass fibers varies depending on the specific type of fiber utilized. However, in general, these fibers are vulnerable to alkalis and moisture when in their bare state, and they can also be prone to stress rust cracking. One characteristic of aramid fibers is their ability to absorb water, which can have an impact on their performance. Additionally, these fibers are prone to significant creep, meaning they may deform over time under constant load. In compression, aramid fibers exhibit nonlinearity, often failing because of kinking and localized buckles after fibrillation (Lau et al., 2016).

When choosing a resin system, designers need to consider various factors such as processing requirements (viscosity, cure temperature, and moisture sensitivities), compatibility with the fibers (sizing or fiber surface therapy), and the desired durability performance (in-service temperatures, toughness, environmental resistance, moisture absorption, UV weathering obstruction, fire resistance).

When it comes to FRP composites, it is important to take into account possible variations in material characteristics, such as long-term performance. One way to address this is by considering partial security or knockdown variables during the design process. Several factors can contribute to uncertainties: a) the amount of uncertainty can arise from the extraction of material attributes either from measurements or models. b) the features of the manufacturing procedure and the degree of cure also play a significant role. c) Additionally, the temperature at which the unit operates and load duration are important factors to consider. It is important to mention that the variability for elastic and strength characteristics tends to be higher for properties that are mainly influenced by the matrix (such as the transverse tensile capacity of a linear lamina) compared to properties that are mainly influenced by the fibers (such as the longitudinal tensile durability of a unidirectional lamina) (Brigante, 2014).

7.5. APPLICATIONS TO THE REHABILITATION OF STRUCTURES

Learning Objectives

- Understand the applications of FRP composites in structural rehabilitation and construction.
- Identify specific scenarios where FRP composites offer advantages in structural rehabilitation.

Composite applications for infrastructure can be categorized into two main areas: structural rehabilitation and new construction, as illustrated in Figure 7.2. In general, FRP composites have the potential to be used for structural recovery in various scenarios (Zhang et al., 2018):

- These include the corrosive effects of deicing salts or exposure to harsh environments.
- Mistakes were made during the building and design phases,
- Alterations to a structure along with shifts in its intended purposes.
- The necessity of fortifying industrial plants and buildings against explosions.
- The necessity of blast-proofing manufacturing operations or buildings
- Design criteria have become more rigorous, especially in terms of seismic resistance to it.
- The aging of materials, as seen in historic buildings.
- Fire damage can have devastating effects (Neale, 2000).
- Vehicle collision

Using composite materials for structural rehabilitation provides numerous benefits compared to their use in new construction. By utilizing fiber-reinforced composites, civil engineers can effectively rehabilitate constructed facilities through methods such as strengthening, repair, or seismic retrofitting:

- Minimize the potential risks associated with implementing a novel material technology.
- Request for a reduced number of designs and construction code provisions.

- Streamlining the process of obtaining construction permits.
- Offer a competitive advantage over traditional materials and techniques (Abbas & Hassan, 2022).

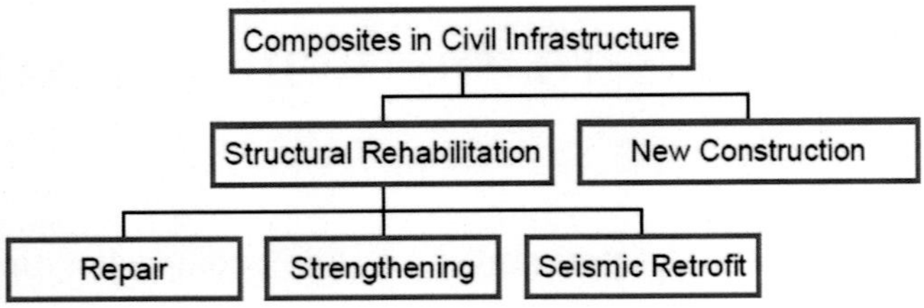

Figure 7.2. *Illustration of classification of composites in civil infrastructure.*

Source: Roberto A. Creative Commons License.

Within the realm of new construction utilizing fiber-reinforced composites, researchers have discovered four primary applications:

- Various methods can be employed to reinforce concrete, such as using rebar, bonded plates, fabrics, or jackets.
- Utilizing prestressed reinforcement within concrete through the use of tendons along with cables;
- Structural shapes commonly employed for columns, beams, and decks;
- Structural systems, like composite shells for columns, are commonly used in various applications.

Table 7.3 indicates four main uses within the field of structural rehabilitation for concrete. Customizing the fiber-reinforced composite structure is necessary to fulfill the specific structural and material demands of each application (Seica & Packer, 2007).

Table 7.3. *Rehabilitation and Retrofit of Concrete Structures with Fiber-Reinforced Composites*

Application	**Description**
I-girder and box beam repair / retrofit	Design for repair: Damage caused by impacts and other incidents during use, as well as damage caused by corrosion; Design for strengthening: Enhancing the load-bearing capacity of structures due to initial design limitations.

Seismic column retrofit	Designing for specific modes of failure includes considerations of shear, lap splice clamping, and bar buckling restraint.
Deck soffit applications	Designing for crack repair along with live load upgrade.
Substructure repairs	Designing for maintenance and strengthening involves considering pile and columns bent caps, pier walls, and non-seismic columns.

There are typically two approaches to rehabilitating and retrofitting existing buildings made of concrete using polymer matrix composites:

- Utilization of composite overlays along with strips.
- Exploring the application of external post-tensioning with composite wires, tendons, or bars.

Composite overlays can be utilized in various scenarios, including the application of full coatings or just the use of strips or patches (Moreillon & Menétrey, 2013).

- Seismic retrofitting and repairing of bridge columns
- Enhancing the strength of the bridge structure (including deck soffits along with girders) to accommodate higher capacity;
- Enhancing the stability and fortification of masonry and concrete walls against seismic activity.
- Several different systems for externally maintaining and retrofitting concrete components were developed, as indicated in Table 7.3. Civil engineers must take into account specific materials problems and worries when applying these systems (Kommajosyula & Rao, 2014).

7.5.1. Rebars

The design protocols found in present standards for buildings made from reinforced concrete are derived from a blend of fundamental engineering concepts and empirical equations. These equations have been developed through rigorous experimental work and decades of practical experience. Regrettably, the current codes solely address steel reinforcement and fail to offer any direction on adapting the existing requirements for utilizing FRP composites in a reinforcement of concrete. Several countries, including Japan, have already developed design processes specifically for utilizing FRP for the reinforcement of concrete (Brühwiler & Denarié, 2013). A series of bridge demonstrations were carried out in Europe, involving thorough evaluation and testing of FRP materials used for concrete reinforcement. The McKinleyville Bridge in West Virginia utilized FRP rods to strengthen the concrete deck supported by steel stringers. In recent years, the ACI Committee 440 released design suggestions for concrete strengthened with FRP bars.

Reinforcement goods for concrete made of constant fibers, such as aramid, carbon, or glass, are impregnated within a polymer matrix. There is a wide range of FRP reinforcement products that come in various forms such as grids, rods, fabrics, and ropes. To achieve a strong reinforcement, it is crucial to establish a solid bond between the FRP material and the concrete (Crasto et al., 2001).

7.5.2. Seismic Retrofit of Bridge Columns

Research has shown that applying an external jacket to concrete can significantly improve its ductility and overall performance during seismic events. This is due to the external jacket's ability to restrict the expansion of concrete when there is not enough hoop steel present. Steel jackets have been widely employed as a seismic retrofit strategy in both Japan and the United States, alongside other conventional methods.

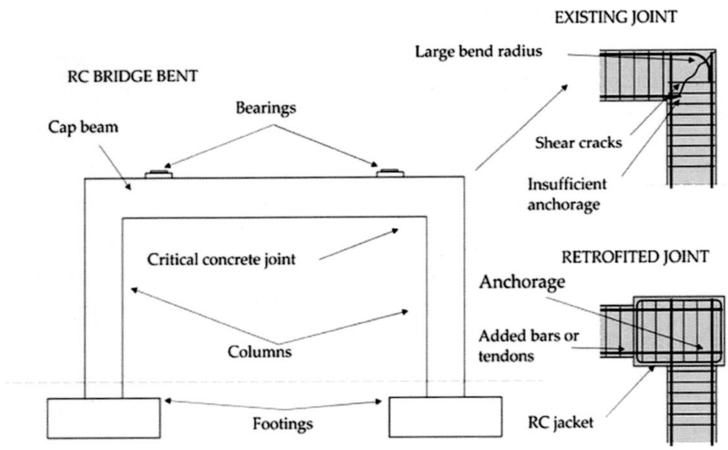

Figure 7.3. *Illustration of the seismic retrofitting of cap beams and concrete joints in RC bridge bents.*

Source: Dominik Skokandić. Creative Commons License.

Nevertheless, the process of retrofitting requires the on-site welding of jacket sections, which raises concerns regarding its efficacy and general monitoring of quality (Tsonos et al., 2017). In certain areas, the utilization of steel may lead to rapid corrosion. Utilizing composites, particularly those reinforced via carbon fibers within the hoop direction allows for a customized layup that optimizes the material's performance for the specific application. This leads to improved material efficiency that surpasses what can be achieved with traditional materials. Research has demonstrated that utilizing composites for retrofitting concrete columns offers a cost-effective alternative to steel retrofits. Additionally, it allows for quicker fabrication, minimizes lane closure, and potentially enhances durability throughout its lifespan. Additional information regarding testing and the selection of materials is documented (Kim & Kim, 2011) (Figure 7.3).

7.5.3. Replacement Bridge Deck Systems

Among the various components of a bridge foundation, bridge decks often demand the highest level of maintenance due to factors such as the wearing surface deteriorating and the deck system degrading over time. In addition to the challenges posed by deterioration, there are also concerns regarding the requirement for greater load ratings (such as HS20 to HS25) and the need to accommodate the growing traffic flow on major roads by adding more lanes. Aside from the financial implications and visible impacts of constantly retrofitting and repairing these structural components, there are also significant consequences in terms of decreased productivity and overall economic setbacks due to delays and detours (Yücel et al., 2022).

Practice Problem

ATLANTEC Constructions' infrastructure division is recognized for its expertise in addressing critical challenges in construction and rehabilitation projects. Specializing in innovative solutions, ATLANTEC is currently faced with a pressing issue concerning an aging concrete bridge deck on a crucial transportation route. Years of exposure to harsh environmental conditions and heavy traffic have resulted in extensive corrosion damage, posing a significant threat to public safety. Traditional replacement materials like concrete and steel present inherent challenges such as susceptibility to corrosion, weight constraints, and prolonged construction timelines, all of which could exacerbate costs and disrupt traffic flow. ATLANTEC Constructions is committed to implementing cutting-edge technologies and sustainable practices to deliver effective and timely solutions, ensuring the continued safety and functionality of vital infrastructure projects.

Solutions to Practice Problem

To address the deterioration of the concrete bridge deck, ATLANTEC Constructions proposes implementing Fiber-Reinforced composites (FRP). FRP offers numerous advantages over conventional materials, including high strength-to-weight ratio, corrosion resistance, and durability. ATLANTEC Constructions engineering team plans to carefully select appropriate FRP composite materials tailored to the specific requirements of the bridge deck, develop a comprehensive design plan ensuring compliance with engineering standards, and utilize efficient installation methods to minimize disruption to traffic. By integrating FRP composites seamlessly into the bridge structure and conducting a thorough cost-benefit analysis, ATLANTEC Constructions aims to deliver a sustainable, cost-effective, and durable replacement solution that ensures the safety of commuters and meets the organization's infrastructure rehabilitation goals.

Therefore, mentioned reasons serve as strong motivation for the advancement of bridge deck materials that possess qualities of durability, light weightiness, and ease of installation. In addition to the potential for reduced maintenance costs, decks made from composites with fiber reinforcement offer significant weight savings. This can lead to savings in substructure expenses and allow for higher actual load ratings in the case of replacing decks. It also opens up the possibility of longer unsustainable spans and improved seismic resistance (Abdelkhalek & Zayed, 2020).

7.5.4. New Structural Systems: Composite Shells

In the field of civil infrastructure, there is significant potential for the use of composites in rehabilitation and retrofit projects. However, there is also great promise in developing new systems that blend the strengths of composite materials via the dominant features of conventional materials, like concrete in compression (Oprisan et al., 2010). These advancements have the potential to greatly enhance the planning and building of new structures for civil engineering. An interesting concept to explore is the use of composite shell frameworks to construct columns. These systems involve the use of prefabricated composite material tubes that serve as both formwork and reinforcement (Gruttmann & Wagner, 2006). By replacing traditional reinforcing steel, these composite tubes allow for faster construction. In this type of structure, the column is built by carefully positioning the hollow composite material shell and subsequently filling it up with concrete. The construction details allow for seamless integration with traditional construction methods (Lund & Stegmann, 2005).

The composite shell method has undergone successful testing using two different approaches. The first approach involves extending starter bars that extend from the foundation into the column, which provides a conventional anchorage system through the footing or bridge superstructure. The second approach eliminates the use of steel entirely and instead relies on the anchorage force mechanism created by the reaction moment resulting from the forces of compression coupled inside the footing, as well as the frictional stresses among the composite shell along with concrete (Totaro & Gürdal, 2009).

SUMMARY

- "Fiber Reinforced Composites in Civil Infrastructure" explores the transformative potential of FRPs in addressing the evolving needs of modern construction. This chapter begins by establishing FRP composites as an emerging class of construction materials, offering innovative solutions to the demands of civil infrastructure.
- The advantages of FRP composites over conventional materials are extensively discussed, including their high specific strength and stiffness, enhanced fatigue life, corrosion resistance, non-magnetic properties, lower life-cycle costs, and faster field installation.
- Despite their numerous benefits, challenges in the adoption of FRP composites as construction materials are identified, such as higher initial material costs, lack of familiarity with materials, and insufficient knowledge of composite structure design
- Furthermore, the chapter delves into the diverse applications of FRP composites in structural rehabilitation, ranging from reinforcing rebar's to seismic retrofit of bridge columns, replacement bridge deck systems, and the implementation of new structural systems like composite shells.
- Overall, the chapter underscores the significant role FRP composites play in revolutionizing civil engineering practices and meeting the evolving needs of infrastructure development.

MULTIPLE CHOICE QUESTIONS

1. **What is a key advantage of fiber-reinforced composites in civil infrastructure?**
   a. High cost
   b. Low stiffness
   c. Corrosion resistance
   d. Magnetic properties
2. **Which property makes FRP composites preferable over conventional materials for bridge decks?**
   a. Low specific strength
   b. High fatigue life
   c. Susceptibility to corrosion
   d. Slow installation process

3. What is a challenge in the adoption of FRP composites as construction materials?

a. Low initial material cost

b. Lack of familiarity with materials

c. Abundance of knowledge on composite structures

d. Uncomplicated design processes

4. Which perspective is crucial for designing FRP composites?

a. Mechanical engineering

b. Materials perspective

c. Biological perspective

d. Historical perspective

5. What is a notable application of FRP composites in structural rehabilitation?

a. Seismic retrofit of bridge columns

b. Increase in magnetic properties

c. Decrease in fatigue life

d. Cost increase in life-cycle

6. Which structural system involves the use of composite shells?

a. Replacement bridge deck systems

b. Seismic retrofit of bridge columns

c. New structural systems

d. None of the above

REVIEW QUESTIONS

1. What are fiber-reinforced composites (FRP) and how do they contribute to civil infrastructure?
2. What advantages do FRP composites offer over conventional construction materials?
3. What challenges are associated with the adoption of FRP composites in construction?
4. How can the higher initial material cost of FRP composites be justified in the long run?
5. What knowledge and skills are essential for effectively designing composite structures using FRP materials?

6. What factors should be considered from a materials perspective when designing with FRP composites?
7. How can FRP composites be applied to rehabilitate structures in civil engineering projects?
8. What innovative structural systems can be developed using FRP composites?

Answers to Multiple Questions

1. (c) 2. (b) 3. (b) 4. (a) 5. (a) 6. (c)

REFERENCES

1. Abbas, H. H., & Hassan, M. M. (2022). Rehabilitation applications of roadway structures. In *Current Perspectives and New Directions in Mechanics, Modelling and Design of Structural Systems* (Vol. 3, pp. 1949–1955). CRC Press.
2. Abdelkhalek, S., & Zayed, T. (2020). Comprehensive inspection system for concrete bridge deck application: Current situation and future needs. *Journal of Performance of Constructed Facilities, 34*(5), 3120001. https://doi.org/10.1061/(ASCE)CF.1943-5509.0001463.
3. Al-Saadi, A. U., Aravinthan, T., & Lokuge, W. (2018). Structural applications of fiber reinforced polymer (FRP) composite tubes: A review of columns members. *Composite Structures, 204*, 513–524. https://doi.org/10.1016/j.compstruct.2018.07.075.
4. Al-Salloum, Y. A., & Almusallam, T. H. (2003). Rehabilitation of the infrastructure using composite materials: Overview and applications. *Journal of King Saud University-Engineering Sciences, 16*(1), 1–20. https://doi.org/10.1016/S1018-3639(09)80043-4.
5. Belarbi, A., Dawood, M., & Acun, B. (2016). Sustainability of fiber-reinforced polymers (FRPs) as a construction material. In *Sustainability of construction materials* (Vol. 3, pp. 521–538). Woodhead Publishing. https://doi.org/10.1016/B978-0-08-100596-8.00027-4.
6. Bobovich, B. B. (2019). Glass-fiber reinforced plastics—Construction materials of the sixth technological paradigm? *Glass and Ceramics, 76*(1), 38–41. https://doi.org/10.1007/s10720-019-00213-1.
7. Brandt, A. M. (2008). Fiber reinforced cement-based (FRC) composites after over 40 years of development in building and civil engineering. *Composite Structures, 86*(1–3), 3–9. https://doi.org/10.1016/j.compstruct.2008.04.009.

8. Brigante, D. (2014). New composite materials. *Nature, 442*, 282–286. https://doi.org/10.1038/442282a.
9. Brühwiler, E., & Denarié, E. (2013). Rehabilitation and strengthening of concrete structures using ultra-high performance fibre reinforced concrete. *Structural Engineering International, 23*(4), 450–457. https://doi.org/10.2749/102232413X13651031743117.
10. Cheng, L., & Karbhari, V. M. (2006). New bridge systems using FRP composites and concrete: A state of the art review. *Progress in Structural Engineering and Materials, 8*(4), 143–154. https://doi.org/10.1002/pse.274.
11. Crasto, A. S., Kim, R. Y., & Mistretta, J. P. (2001). The application of composites for the rehabilitation of concrete bridge infrastructure. *Advanced Composite Materials, 10*(2–3), 147–157. https://doi.org/10.1163/15685510152761030.
12. Das, S. C., & Nizam, M. (2014). Applications of fiber reinforced polymer composites (FRP) in civil engineering. *International Journal of Advanced Structures and Geotechnical Engineering, 3*(3), 299–309. https://doi.org/10.6088/ijasege.2014030204.
13. Frigione, M., & Lettieri, M. (2018). Durability issues and challenges for material advancements in FRP employed in the construction industry. *Polymers, 10*(3), 247. https://doi.org/10.3390/polym10030247.
14. Gruttmann, F., & Wagner, W. (2006). Structural analysis of composite laminates using a mixed hybrid shell element. *Computational Mechanics, 37*, 479–497. https://doi.org/10.1007/s00466-005-0028-7.
15. Gudonis, E., Timinskas, E., Gribniak, V., Kaklauskas, G., Arnautov, A. K., & Tamulėnas, V. (2013). FRP reinforcement for concrete structures: State-of-the-art review of application and design. *Engineering Structures and Technologies, 5*(4), 147–158. https://doi.org/10.3846/20297955.2013.868321.
16. Harris, B. (1991). A perspective view of composite materials development. *Materials & Design, 12*(5), 259–272. https://doi.org/10.1016/0261-3069(91)90063-7.
17. Hollaway, L. C. (2003). The evolution of and the way forward for advanced polymer composites in the civil infrastructure. *Construction and Building Materials, 17*(6–7), 365–378. https://doi.org/10.1016/S0950-0618(03)00041-8.
18. Karbhari, V. M., Chin, J., Hunston, D., Benmokrane, B., Juska, T., Morgan, R., & Reynaud, A. D. (2003). Durability gap analysis for fiber-reinforced polymer composites in civil infrastructure. *Journal of Composites for*

Construction, 7(3), 238–247. https://doi.org/10.1061/(ASCE)1090-0268(2003)7:3(238).

19. Khalid, M. Y., Al Rashid, A., Arif, Z. U., Ahmed, W., Arshad, H., & Zaidi, A. A. (2021). Natural fiber reinforced composites: Sustainable materials for emerging applications. *Results in Engineering, 11*, 100263. https://doi.org/10.1016/j.rineng.2021.100263.
20. Kim, S. H., & Kim, D. K. (2011). Seismic retrofit of rectangular RC bridge columns using wire mesh wrap casing. *KSCE Journal of Civil Engineering, 15*, 1227–1236. https://doi.org/10.1007/s12205-011-1227-4.
21. Kommajosyula, R., & Rao, N. V. R. (2014). Innovative applications of post-tension for restoration and rehabilitation of bridges and structures—Case studies in India. *Journal of Civil Engineering and Architecture, 8*(7), 741–745. https://doi.org/10.17265/1934-7359/2014.07.007.
22. Kossakowski, P. G., & Wciślik, W. (2022). Fiber-reinforced polymer composites in the construction of bridges: Opportunities, problems and challenges. *Fibers, 10*(4), 37. https://doi.org/10.3390/fib10040037.
23. Lau, D., Qiu, Q., Zhou, A., & Chow, C. L. (2016). Long term performance and fire safety aspect of FRP composites used in building structures. *Construction and Building Materials, 126*, 573–585. https://doi.org/10.1016/j.conbuildmat.2016.09.072.
24. Mirmiran, A. (2004). Bonded repair and retrofit of concrete structures using FRP composites: Recommended construction specifications and process control manual. *Transportation Research Board* (Vol. 514, pp. 741–745).
25. Naser, M. Z., Hawileh, R. A., & Abdalla, J. A. (2019). Fiber-reinforced polymer composites in strengthening reinforced concrete structures: A critical review. *Engineering Structures, 198*, 109542. https://doi.org/10.1016/j.engstruct.2019.109542.
26. Neale, K. W. (2000). FRPs for structural rehabilitation: A survey of recent progress. *Progress in Structural Engineering and Materials, 2*(2), 133–138. https://doi.org/10.1002/pse.63.
27. Noël, M. (2019). Probabilistic fatigue life modelling of FRP composites for construction. *Construction and Building Materials, 206*, 279–286. https://doi.org/10.1016/j.conbuildmat.2019.01.090.
28. Oprişan, G., Ţăranu, N., Munteanu, V., & Enţuc, I. (2010). Application of modern polymeric composite materials in industrial construction. *Bulletin of the Polytechnic Institute of Iaşi, 56*, 121–130. https://doi.org/10.1515/bpi-2015-0032.

29. Orcesi, A. D., Feraille, A., & Chataigner, S. (2019). Fatigue strengthening of steel structures using high modulus CFRP plates: Development of a life-cycle analysis approach. *Construction and Building Materials, 227*, 116628. https://doi.org/10.1016/j.conbuildmat.2019.116628.
30. Porter, M. L., & Harries, K. (2007). Future directions for research in FRP composites in concrete construction. *Journal of Composites for Construction, 11*(3), 252–257. https://doi.org/10.1061/(ASCE)1090-0268(2007)11:3(252).
31. Seica, M. V., & Packer, J. A. (2007). FRP materials for the rehabilitation of tubular steel structures, for underwater applications. *Composite Structures, 80*(3), 440–450. https://doi.org/10.1016/j.compstruct.2006.05.003.
32. Sharma, A., & Joshi, S. C. (2023). Enhancement in fatigue performance of FRP composites with various fillers: A review. *Composite Structures, 309*, 116724. https://doi.org/10.1016/j.compstruct.2023.116724.
33. Totaro, G., & Gürdal, Z. (2009). Optimal design of composite lattice shell structures for aerospace applications. *Aerospace Science and Technology, 13*(4–5), 157–164. https://doi.org/10.1016/j.ast.2008.11.001.
34. Tsonos, A. D. G., Kalogeropoulos, G. I., Iakovidis, P. E., & Konstantinidis, D. (2017). Seismic retrofitting of pre-1970 RC bridge columns using innovative jackets. *International Journal of Structural Engineering, 8*(2), 133–147. https://doi.org/10.1504/IJSTRUCTE.2017.085443.
35. Uomoto, T., Mutsuyoshi, H., Katsuki, F., & Misra, S. (2002). Use of fiber reinforced polymer composites as reinforcing material for concrete. *Journal of Materials in Civil Engineering, 14*(3), 191–209. https://doi.org/10.1061/(ASCE)0899-1561(2002)14:3(191).
36. Yücel, H. E., Dutkiewicz, M., & Yıldızhan, F. (2022). Application of ECC as a repair/retrofit and pavement/bridge deck material for sustainable structures: A review. *Materials, 15*(24), 8752. https://doi.org/10.3390/ma15248752.
37. Zaman, A., Gutub, S. A., & Wafa, M. A. (2013). A review on FRP composites applications and durability concerns in the construction sector. *Journal of Reinforced Plastics and Composites, 32*(24), 1966–1988. https://doi.org/10.1177/0731684413501402.
38. Zhou, X. Y., Qian, S. Y., Wang, N. W., Xiong, W., & Wu, W. Q. (2022). A review on stochastic multiscale analysis for FRP composite structures. *Composite Structures, 284*, 115132. https://doi.org/10.1016/j.compstruct.2022.115132.

CHAPTER 8

EXPLORING ADVANCED STEEL APPLICATIONS IN CIVIL ENGINEERING

LEARNING OBJECTIVES

At the end of this chapter, readers will be able to:

1. Understand the historical evolution of steel in civil engineering practice.
2. Identify key issues concerning fire protection, buckling, and fatigue in steel structures.
3. Explore advancements in steel alloys, such as High-Performance Steel (HPS) in bridge construction, and their applications.
4. Examine innovative steel structures like buckling restrained braces and concrete filled steel tubes.
5. Analyze the benefits and challenges of composite and hybrid systems in steel construction.
6. Discuss the significance of corrosion resistance and weldability in modern steel engineering.

INTRODUCTORY EXAMPLE

Zeta Engineering, a globally acclaimed firm revered for its pioneering contributions to civil engineering, recently tackled the ambitious project of building a skyscraper in a seismic-prone region. This endeavor presented an array of challenges, from ensuring structural integrity to optimizing space and minimizing costs.

In response, Zeta Engineering embarked on a journey of innovation, leveraging advanced steel applications to redefine modern construction practices. By incorporating state-of-the-art methodologies and materials, such as Buckling Restrained Braces (BRBs) and Concrete Filled Steel Tubes (CFT), the firm successfully engineered a skyscraper that transcended traditional boundaries. Through meticulous planning and strategic implementation, Zeta Engineering fortified the building against seismic forces while maximizing interior space and streamlining the construction process. The completed skyscraper stands as a towering testament to Zeta Engineering's unwavering commitment to pushing the envelope of steel innovation, setting new benchmarks of excellence in urban infrastructure worldwide.

UNIT INTRODUCTION

Steel has been a popular choice for construction in civil engineering systems for over a century. Before the late 19th century, civil engineering structures utilized cast and wrought iron components for columns and other elements. However, these structural elements often exhibited unpredictable failures, limited flexibility, and consistent strength in comparison to the more advanced steel structures of today. In the year 1890, iron-cast and iron-wrought columns fell out of favor, and structural steel emerged as the preferred construction material (Pardeshi & Patil, 2021). This happened due to advancements in the Bessemer process, which allowed for cost-effective production of the significant amounts of steel needed for constructing buildings and bridges. Structural steel offers consistent material behavior, exceptional strength, and remarkable ductility. The growing use of steel in construction was a result of its exceptional properties. However, the early development of steel structures subsequently followed the methods of building used for iron-cast and iron-wrought elements. Standard codes along with specifications were not readily accessible until the 1920s, and as a result, early structures were built based on established practices and the lessons learned from previous construction projects (Liu, 2011).

Expert structural engineers crafted buildings and bridges using straightforward calculations along with stress levels which had developed in iron-cast and iron-wrought iron components. The cost of structural steel was quite high, whereas the cost of work was comparatively low. As a result, steel was manufactured in the form of plates, bars, and angles. Many structural elements were constructed using these limited shapes, which were assembled into built-up members, as shown in Figure 8.1 (Sarkar et al., 2024).

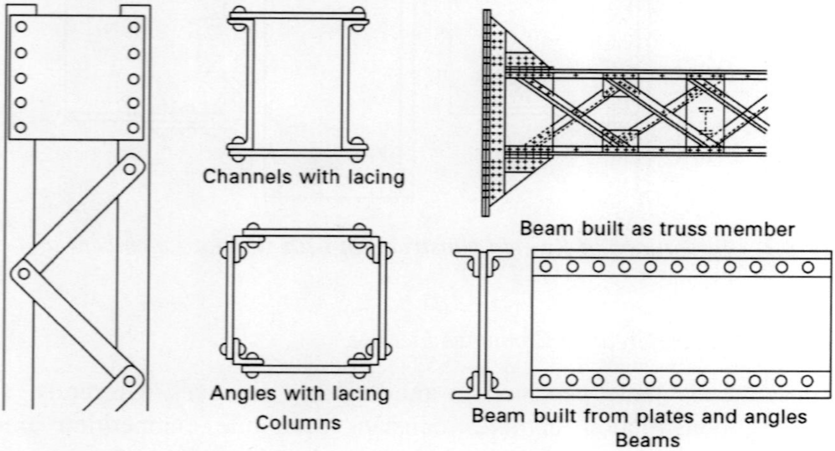

Figure 8.1. *Illustration of Historic steel construction.*

Source: Daniel P, Creative Commons License.

In the 1920s, there was a notable shift in steel design practice across many parts of the world, primarily due to rising labor costs. Various factors, such as local traditions and financial circumstances, have contributed to the delayed adoption of these advancements in certain countries. As a result, remnants of earlier practices can still be observed in a few nations even today. Design specifications for buildings as well as bridges were initially developed, leading to the evolution of uniformed hot-rolled shapes that are frequently utilized in modern construction, such as wide flanges and channels (Amirtharaj & Mariappan, 2023). Conventional connections, like the ones shown in Figure 8.2, were utilized due to their reduced labor requirements and the use of fewer rivets compared to previous construction methods. The initial specifications utilized the permitted stress method, with permissible stress levels of around 125 MPa for various structural applications. Common structural shapes like I beams, H beams, and channels have significantly decreased the amount of labor needed for constructing the building. In addition, advancements in the manufacturing method of steel have led to cost savings and a greater utilization of this material (Irizarry et al., 2012).

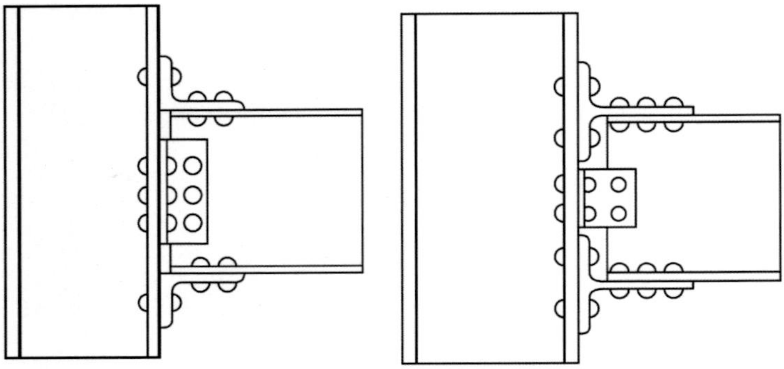

Figure 8.2. *Illustrations of Riveted construction with standard members and connections.*

Source: C W Roder, Creative Commons License.

These trends have persisted in modern-day practices. Currently, civil engineering construction involves constant economic competition among different construction materials, including steel and concrete. Consequently, there have been consistent advancements in the properties of these materials. Currently, the typical mild steels utilized in civil engineering construction boast nominal yield stress values of around 345 MPa (Banhart, 2007). The allowable stress values for these kinds of steels typically range from 205 to 220 MPa. The operating stress levels have significantly increased compared to when steel initially was used. Higher levels of operating stress can lead to an increased risk of stability failures, fatigue, and brittle fractures. As a result, structural engineers need to address these design challenges, and building codes or specifications are getting increasingly intricate. With the rising labor costs, engineers are now challenged to design structures that are lightweight, cost-effective, and can be assembled efficiently (Khan et al., 2022).

Some of these kinds of uses are in challenging environments, such as salt water, where the effects of rust, protection from corrosion, and degradation are of growing concern. Steel is the preferred choice for larger, taller, and more extensive civil engineering structures due to its superior material durability, remarkable durability-to-weight ratio, and excellent ductility. Steel is widely used in the creation and erection of large structures and those with unique shapes due to its remarkable durability-to-weight ratio and cost-effective fabrication and installation of complicated frameworks. The industry of steel is constantly striving to enhance material performance, increase strength, and develop more flexible materials and structures, as these qualities contribute to its ongoing progress (Li et al., 2016).

8.1. ISSUES OF CONCERN

Learning Objectives

- Understand the key concerns in steel structures: fire protection, buckling, fatigue, fracture, corrosion, and weldability.
- Know the impact of factors like material strength and ductility on seismic design.

The previous analysis of the development of steel usage in the civil engineering field brings up several important considerations for engineering conception, research, and the advancement of steel structures. Some of the problems that arise are (Leite et al., 2016):

8.1.1. Fire Protection

Significant progress has been achieved in developing more efficient and cost-effective fire protection techniques. However, fire protection remains an important issue for steel structures due to their smaller members and lower thermal mass compared to other building materials. Fire protection is a crucial consideration in civil engineering projects, as it adds to the overall structural cost. Engineers are constantly striving to find ways to minimize these costs without compromising the ability of the structure to withstand high temperatures during a fire (Gernay & Khorasani, 2020).

8.1.2. Buckling

Steel structures typically feature smaller members and higher operating stress levels compared to alternative construction materials. As more powerful materials continue to advance, the importance of buckling and stability becomes increasingly critical.

8.1.3. Fatigue and Fracture

Crack initiation, crack growth, and propagation have gained major significance in structural engineering over the past few decades. Nevertheless, the growing yield stress, operational stress levels, focus on plastic or ultimate capacity, and adoption of welded building process have led to a rise in the occurrence of fatigue and fracture within bridges and manufacturing processes. Currently, fatigue is not a significant consideration in building design. However, the cracking of steel in construction has emerged as a prominent concern for seismic design following recent earthquakes (Fisher & Roy, 2011).

8.1.4. Corrosion Resistance

Corrosion of steel when subjected to the environment can result in deterioration, heightened maintenance requirements, and escalated construction expenses. Methods such as galvanization, paint, and coatings can effectively prevent corrosion, although they may slightly raise the fabrication expenses of the steel. As a result, engineers are always looking for ways to lower these expenses (Wang et al., 2022).

8.1.5. Weldability

Welded construction is a popular choice in civil engineering due to its ability to create structures that are more rigid and robust, while also minimizing the weight and amount of materials needed. The growing strength of steel presents challenges for welding techniques, as higher-strength steels can be more challenging to weld without compromising the system's ductility and performance. Enhanced structural flexibility for earthquakes and other types of loading. Seismic design has become a necessity for numerous structural systems in civil engineering. Steel, with its exceptional material strength, stiffness, and ductility, emerges as an ideal choice for seismic design. Factors such as buckling and weldability can have an impact on the seismic resilience and ductility of structures. Engineers are constantly working on innovative approaches to enhance the inelastic seismic behavior of steel structures.

The decision and consequences of employing steel as a building component are influenced by various factors. These factors often drive developments aimed at enhancing the performance of steel (David et al., 2013).

8.2. INNOVATIONS IN STEEL

Learning Objectives

- Understand the development and benefits of High-Performance Steels (HPS) for bridge construction.
- Explore the characteristics and applications of fire-resistant steel alloys.

Current challenges and issues often drive the development of new engineering solutions. Several recent improvements and advancements in steel have been made in response to the frequent problems mentioned earlier. This article explores several recent advancements in the field, focusing on the progress made in steel alloys, components, and structural systems. These advancements have resulted in a construction material that is more versatile and cost-effective (Birat, 2018).

8.2.1. Innovative Steel Alloys

In recent years, there have been advancements in steel alloys that offer enhanced structural performance and cost-effectiveness for steel construction projects. Overall, these alloys help to resolve the concerns mentioned earlier. Here, some of the material developments are discussed (Kisku, 2020).

8.2.1.1. HPS Bridge Steel

Advanced steels with enhanced properties have been created to meet the demands of modern bridge construction, offering increased yield strength, weldability, ductility, and fatigue resistance. HPS steels are quenched and tempered lightweight plate steels produced under ASTM 709. They were specifically designed to offer excellent ductility, toughness, and weldability, even at high-yield stress levels. HPS steels enable the creation of lighter bridges, have longer spans, and are more cost-effective to construct. Nevertheless, the increased operating stress levels necessitate modifications in the construction of bridge practice. Higher stress levels lead to the need for more compact flanges and webs, which in turn require stronger elements to fully utilize their plastic capacity. Extensive research has been conducted on the HPS70 alloy to assess its strength and ductility. As a result, new design limits have been suggested to optimize the use of this high-strength steel in bridge design, ensuring cost-effective construction (Earls & Shah, 2002).

8.2.1.2. Fire-resistant Steels

Fire resistance is an important factor to consider when evaluating the performance of steel under high temperatures. It is generally accepted that steel is said to be fire-resistant if it can maintain 0.6–0.7 of its strength at room temperature when exposed to high temperatures over short periods. Steels with the Cr-Mo-V-Nb alloying system can withstand high temperatures, making them suitable for fire-resistant applications up to 700°C. As shown in Figure 8.3, Young's modulus of steel maintains a consistent level as the temperature rises (from room temperature to about 700°C). However, in both of them, the yield or tensile force of steel gradually decreases as the temperature increases. It is important to note that buildings can be at risk of collapsing during fires. Therefore, the fire resistance of steel and the fire-resistant properties of steel are critical factors to consider in structural design (Sha et al., 2001).

8.2.1.3. Corrosion Resistant Steels

One drawback of steel is its susceptibility to corrosion when subjected to the environment. This has been a major economic concern and sometimes discourages the use of steel in structural systems that are exposed to the elements. Steels that are resistant to corrosion are created by incorporating over 12% chromium, leading to the development of a thin, corrosion-resistant, and inert oxide film. An oxide film can form and repair itself when exposed to oxygen (Bardal et al., 1993).

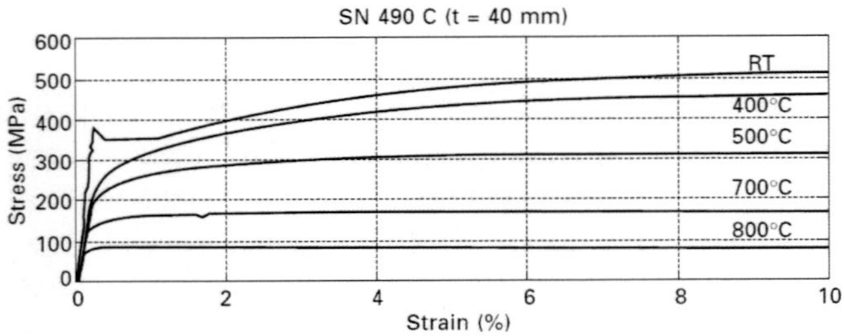

Figure 8.3. *Description of the effects of temperature on a material's stress–strain relationship.*

Source: M Nakshima, Creative Commons License.

8.3. INNOVATIVE STEEL STRUCTURES

Learning Objectives

- Understand the function and advantages of Buckling Restrained Braces (BRBs) in structural engineering.
- Explore the benefits and applications of Concrete Filled Steel Tubes (CFT) in modern construction.
- Know the principles and design considerations of Additional Composite and Hybrid Systems in building construction.

Steel, a material known for its robustness and adaptability, is constantly being utilized in innovative ways in the field of construction and engineering. Architects and engineers are constantly pushing the limits of steel, exploring advanced alloys and innovative structural designs. This section delves into the various ways in which steel is being used to improve the efficiency and long-term viability of buildings and infrastructure. It highlights the innovative applications of steel in these areas (Kanno, 2016).

8.3.1. Buckling Restrained Braces

BRB (a buckling-restrained brace) has become a type of structural brace commonly used in buildings to enhance their ability to withstand periodic lateral loadings, particularly those caused by earthquakes. The structure comprises a slender steel core, a concrete enclosure that provides continuous support to the core and prevents buckling under axial stress, and a junction region that avoids any undesired interactions among the two components. Braced frames that utilize BRBs, referred to as buckling-restrained braced frames or BRBFs, offer notable benefits compared to conventional braced frames (Dong et al., 2017) (Figure 8.4).

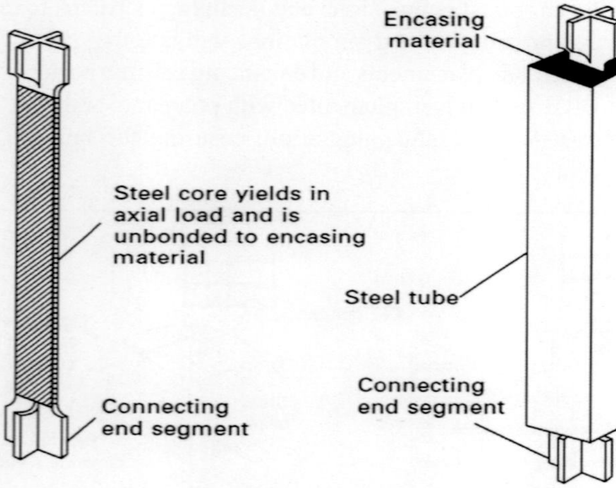

Figure 8.4. *Schematic of buckling restrained brace.*

Source: M Nakshima, Creative Commons License.

Practice Problem

SANVY Construction Company is a leading firm specializing in the design and construction of high-rise buildings, particularly in seismically active regions. With a strong commitment to safety and innovation, SANVY's engineering team is currently focused on developing robust structural systems capable of withstanding significant seismic loads. As they undertake the design of a new high-rise building, ensuring the safety of occupants remains paramount. SANVY Construction is dedicated to implementing state-of-the-art solutions that enhance the building's resilience against lateral forces, particularly during seismic events. By integrating advanced engineering principles and cutting-edge technology, SANVY Construction strives to deliver structures that not only meet stringent safety standards but also exemplify excellence in seismic-resistant design.

Solutions to Practice Problem

SANVY Construction Company integrates buckling restrained braces (BRBs) into the structural framework of the high-rise building, selecting suitable BRBs based on their capacity to resist lateral loads and compatibility with the building's structural system. Through meticulous planning, they identify optimal locations for BRB installation and design robust connections and anchorage systems to ensure proper load transfer. Structural analysis and simulations validate the effectiveness of the BRBs in reducing lateral displacements and enhancing seismic performance. During construction, the BRB system is implemented with precision, and post-construction monitoring confirms its successful integration, ensuring the building's resilience against seismic events.

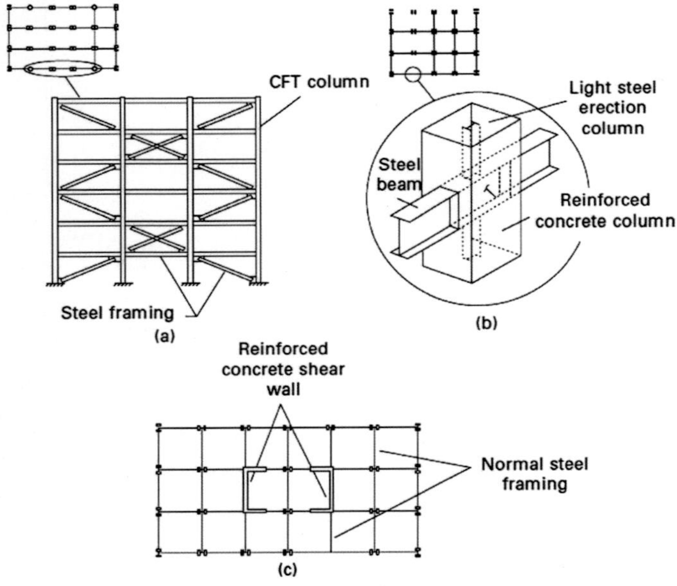

Figure 8.5. *Illustration of New development in composite and hybrid structural systems; (a) concrete filled steel tube; (b) RCS system; and (c) hybrid RC shear wall and steel frame system.*

Source: J Woodward, Creative Commons License.

8.3.2. Concrete Filled Steel Tubes (CFT)

Composite or hybrid uses in civil engineering often yield the greatest benefits when utilizing structural steel. By combining concrete with steel, substantial economic and durability advantages can be achieved. Structural steel, on the other hand, offers a greater compressive strength in comparison to concrete. This allows for a more efficient use of materials, as less volume of steel must be used to support compressive loads. This results in improved structure stiffness and durability against buckling. Concrete possesses limited shear strength, virtually no tensile strength, and limited ductility. Steel demonstrates exceptional capabilities in these fields, and the resultant composite and hybrid elements often possess superior strength, ductility, and cost-effectiveness compared to standalone steel or concrete-reinforced elements (Milan et al., 2019).

The combination of steel and concrete offers numerous possibilities for effective structural design. One particularly popular choice is the use of concrete-filled steel tubes (CFT), as shown in Figure 8.5, which is gaining popularity in modern construction. Utilizing high-strength steel tubes allows for the incorporation of smaller and lighter members that offer enhanced resistance to buckling, stiffness, and stability. The use of CFT can lead to cost savings in construction due to the dual function of the tube as both structure and reinforcement for the concrete. This can help reduce labor costs associated with reinforcing the concrete element (Tiwary, 2022). The CFT member is significantly lighter compared to the reinforced concrete needed for the same load and deformation, resulting in a further reduction of seismic design forces and design dead loads. The addition of concrete fill enhances the compressive stiffness of the structure, thereby increasing the axial stiffness of the column and preventing buckling of the steel. CFT columns are highly sought after in as super columns as collector columns or in braced frames within taller buildings, due to their enhanced axial column compressive strength and stiffness. The use of concrete fill allows for greater ductility by using thinner and more slender tubes, surpassing what steel alone can achieve. The tube's confining action enhances the shear strength and inelastic strain along with the ability of the concrete, thereby boosting the overall ductility of the system (Chang et al., 2014).

8.3.3. Additional Composite and Hybrid Systems

New forms of material composite or hybrid building have been developed lately. The advancement of a hybrid reinforced concrete-steel reinforced concrete (RCS) construction, as shown in Figure 8.6(b), was first initiated in the United States over 15 years ago. Today, this construction method is widely utilized in both the United States and Japan for intermediate-sized buildings constructed with moment frame construction (Abeysinghe et al., 2013).

Typically, RCS construction begins by assembling a steel frame, using either light columns made of steel or steel columns that are spaced further apart. Moment-resisting frames are crucial in civil engineering as they play a vital role in withstanding lateral loads such as wind and earthquakes. This is achieved by effectively transferring bending moment and shear forces from beams to columns. The columns, with their significant bending resistance and stiffness, contribute to the overall structural integrity. During seismic design, it is anticipated that the beams near the column will experience inelastic deformation. Thus, the beam-column connection must possess sufficient strength to effectively transmit these forces and moments, while also ensuring that the beam's ability to undergo inelastic deformation is not compromised.

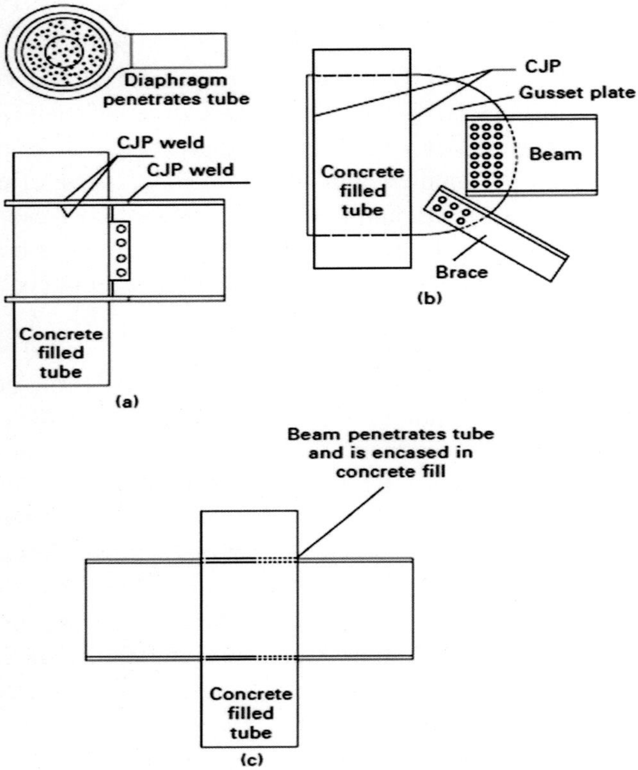

Figure 8.6. *Illustration of the Connections used for CFT construction; (a) internal diaphragm moment frame connection; (b) penetrating gusset plate CBF connection; and (c) penetrating beam moment frame connection.*

Source: David M, Creative Commons License.

Additionally, the column must exhibit the necessary stiffness and strength to regulate frame deflections and generate the desired lateral resistance for the structure. Reinforced concrete (RC) columns offer cost-effective advantages when incorporated into steel frame construction in RCS projects. The RC columns are poured among widely separated steel columns, or the RC column could protect the lightweight steel columns that were utilized for supporting the steel framing during erection. The reinforced concrete column includes the steel beam to create a moment-resisting frame, as shown in the figure above (Loss et al., 2016).

The RC column can be constructed to provide significant flexural stiffness and obstruction, but the design details of the connections are crucial for the design of the RC Column System. Proper connection design plays a crucial role as it deals with significant compressive bearing stresses that occur at the steel-concrete interactions within the connection. Additionally, the shear design of the link takes into account the shear strength of the two materials steel and concrete (Deierlein, 2000).

In recent years, there has been an increasing use of hybrid systems, such as the combination of RC shear walls and steel framing, as shown in Figure 8.6(c). Previous versions of these alternatives have been utilized in previous years. However, modern construction methods often involve the use of slip or jumping-creating techniques to build the shear wall (Hoffmeister et al., 2017).

SUMMARY

- In this chapter, we delved into the realm of advanced steel applications within civil engineering, uncovering the myriad ways in which steel innovations are reshaping the infrastructure landscape.
- We began by addressing critical issues of concern such as fire protection, buckling, fatigue and fracture, corrosion resistance, and weldability, highlighting the pivotal role of advanced steel in mitigating these challenges.
- Moving forward, we explored innovations in steel, focusing on the development of innovative steel alloys like HPS for Bridges, fire-resistant steels, and corrosion-resistant steels. These advancements underscored the continuous evolution of steel to meet the demands of modern engineering practices, offering enhanced structural performance and durability.
- Finally, we examined innovative steel structures, including buckling restrained braces, concrete-filled steel tubes (CFT), and additional composite and hybrid systems.
- These cutting-edge applications exemplify the versatility and adaptability of steel in creating resilient and sustainable infrastructure solutions.

MULTIPLE CHOICE QUESTIONS

1. **What is a primary concern addressed by fire-resistant steel in civil engineering?**
   a. Corrosion protection
   b. Enhanced ductility
   c. Weldability
   d. Structural stability
2. **Buckling restrained braces are primarily utilized in buildings for:**
   a. Withstanding lateral loadings
   b. Increasing weldability
   c. Enhancing corrosion resistance
   d. Improving fatigue resistance
3. **What advantage do Concrete Filled Steel Tubes (CFT) offer in structural design?**
   a. Reduced weldability
   b. Increased corrosion

c. Enhanced ductility
d. Improved compressive stiffness

4. What is the primary purpose of innovative steel alloys such as HPS Bridge Steels?

a. Improved corrosion resistance
b. Increased weldability
c. Enhanced structural performance
d. Higher ductility

5. Additional Composite and Hybrid Systems, like RCS construction, typically involve:

a. Exclusive use of steel
b. Combining reinforced concrete with steel
c. Incorporating only fire-resistant steels
d. Focusing solely on weldability improvements

REVIEW QUESTIONS

1. What are the primary concerns addressed by fire-resistant steel in civil engineering?
2. Explain the significance of buckling restrained braces in modern structural engineering.
3. How do concrete filled steel tubes offer advantages in structural design?
4. Discuss the importance of corrosion resistance and its impact on steel construction.
5. Analyze the advancements in steel alloys, such as HPS Bridge Steels, and their applications.
6. Explore the benefits and challenges of composite and hybrid systems in steel construction.

Answers to Multiple Questions

1. (d) 2. (a) 3. (d) 4. (c) 5. (b)

REFERENCES

1. Abeysinghe, C. M., Thambiratnam, D. P., & Perera, N. J. (2013). Flexural performance of an innovative hybrid composite floor plate

system comprising glass–fiber reinforced cement, polyurethane and steel laminate. *Composite Structures, 95*, 179–190. https://doi.org/10.1016/j.compstruct.2012.07.027.

2. Amirtharaj, J., & Mariappan, M. (2023). Exploring the potential uses of Aluminum Metal Matrix Composites (AMMCs) as alternatives to steel bar in Reinforced Concrete (RC) structures: A state-of-art review. *Journal of Building Engineering, 4*(4), 108085. https://doi.org/10.1016/j.jobe.2023.108085.
3. Banhart, J. (2007). Metal foams—From fundamental research to applications. *Frontiers in the Design of Materials, 7*(2), 279–285. https://doi.org/10.1007/s11940-007-0020-5.
4. Bardal, E., Drugli, J. M., & Gartland, P. O. (1993). The behaviour of corrosion-resistant steels in seawater: A review. *Corrosion Science, 35*(1–4), 257–267. https://doi.org/10.1016/0010-938X(93)90063-4.
5. Birat, J. P. (2018). Product innovations of key economic importance for the steel industry. *Metallurgical Research & Technology, 115*(4), 420. https://doi.org/10.1051/metal/2018045.
6. Chang, X., Luo, X., Zhu, C., & Tang, C. (2014). Analysis of circular concrete-filled steel tube (CFT) support in high ground stress conditions. *Tunnelling and Underground Space Technology, 43*, 41–48. https://doi.org/10.1016/j.tust.2014.01.004.
7. David, S. A., Siefert, J. A., & Feng, Z. (2013). Welding and weldability of candidate ferritic alloys for future advanced ultra supercritical fossil power plants. *Science and Technology of Welding and Joining, 18*(8), 631–651. https://doi.org/10.1179/1362171813Y.0000000155.
8. Deierlein, G. G. (2000). New provisions for the seismic design of composite and hybrid structures. *Earthquake Spectra, 16*(1), 163–178. https://doi.org/10.1193/1.1586128.
9. Dong, H., Du, X., Han, Q., Hao, H., Bi, K., & Wang, X. (2017). Performance of an innovative self-centering buckling restrained brace for mitigating seismic responses of bridge structures with double-column piers. *Engineering Structures, 148*, 47–62. https://doi.org/10.1016/j.engstruct.2017.05.009.
10. Earls, C. J., & Shah, B. J. (2002). High performance steel bridge girder compactness. *Journal of Constructional Steel Research, 58*(5–8), 859–880. https://doi.org/10.1016/S0143-974X(02)00004-7.
11. Fisher, J. W., & Roy, S. (2011). Fatigue of steel bridge infrastructure. *Structure and Infrastructure Engineering, 7*(7–8), 457–475. https://doi.org/10.1080/15732479.2010.539441.

12. Kanno, R. (2016). Advances in steel materials for innovative and elegant steel structures in Japan: A review. *Structural Engineering International, 26*(3), 242–253. https://doi.org/10.2749/101686616X14575991380316.
13. Kisku, N. (2020). Strengthening of high-alloy steel through innovative heat treatment routes. *Welding-Modern Topics, 7*(3), 12–45. https://doi.org/10.21795/wmt.2020.12.
14. Li, Y., Gao, Z., & Zhang, X. (2016). BIM technology application in the exploration of super high-rise office building project in civil construction. In *2016 International Conference on Civil, Transportation and Environment* (Vol. 2, pp. 269–272). Atlantis Press.
15. Liu, G. (2011). Steels: Data exploration for discovery and data-sharing. In *Advanced Steels: The Recent Scenario in Steel Science and Technology* (Vol. 2, pp. 131–136). Springer Berlin Heidelberg. https://doi.org/10.1007/978-3-642-18152-3_11.
16. Loss, C., Piazza, M., & Zandonini, R. (2016). Connections for steel–timber hybrid prefabricated buildings. Part II: Innovative modular structures. *Construction and Building Materials, 122*, 796–808. https://doi.org/10.1016/j.conbuildmat.2016.06.054.
17. Milan, C. C., Albareda-Valls, A., & Carreras, J. M. (2019). Evaluation of structural performance between active and passive preloading systems in circular concrete-filled steel tubes (CFST). *Engineering Structures, 194*, 207–219. https://doi.org/10.1016/j.engstruct.2019.05.059.
18. Pardeshi, R. T., & Patil, Y. D. (2021). Review of various shear connectors in composite structures. *Advanced Steel Construction, 17*(4), 394–402. https://doi.org/10.18057/ASC2021.17.4.3.
19. Sarkar, K., Shiuly, A., & Dhal, K. G. (2024). Revolutionizing concrete analysis: An in-depth survey of AI-powered insights with image-centric approaches on comprehensive quality control, advanced crack detection, and concrete property exploration. *Construction and Building Materials, 411*, 134212. https://doi.org/10.1016/j.conbuildmat.2023.134212.
20. Sha, W., Kirby, B. R., & Kelly, F. S. (2001). The behaviour of structural steels at elevated temperatures and the design of fire-resistant steels. *Materials Transactions, 42*(9), 1913–1927. https://doi.org/10.2320/matertrans1998.42.1913.
21. Tiwary, A. K. (2022). Experimental investigation into mild steel circular concrete-filled double skin steel tube columns. *Journal of Constructional Steel Research, 198*, 107527. https://doi.org/10.1016/j.jcsr.2022.107527.
22. Wang, N., Wang, Q., & Xu, S. (2022). A review on applications of superhydrophobic materials in civil engineering. *Advanced Engineering Materials, 24*(6), 2101238. https://doi.org/10.1002/adem.202101238.

INDEX

Q

R

S